SpringerBriefs in Microbiology

More information about this series at http://www.springer.com/series/8911

Daniele Focosi

SARS-CoV-2 Spike Protein Convergent Evolution

Impact of Virus Variants on Efficacy of COVID-19 Therapeutics and Vaccines

 Springer

Daniele Focosi (iD)
North-Western Tuscany Blood Bank
Pisa University Hospital
Azienda Ospedaliera Universitaria Pisana
Pisa, Italy

ISSN 2191-5385 ISSN 2191-5393 (electronic)
SpringerBriefs in Microbiology
ISBN 978-3-030-87323-3 ISBN 978-3-030-87324-0 (eBook)
https://doi.org/10.1007/978-3-030-87324-0

This Springer imprint is published by the registered company Springer Nature Switzerland AG
The registered company address is: Gewerbestrasse 11, 6330 Cham, Switzerland

"Everything existing in the universe is the fruit of chance and necessity" (Democritus)

To my only loves, my wife Emona and my children Enea and Anna.

About This Book

The Spike protein represents the target of vaccines and antibody-based therapeutics (convalescent plasma, hyperimmune sera and monoclonal antibodies) for COVID-19. Spike mutations can affect the efficacy of those treatments. Continuous monitoring of such mutations is necessary to reshape the inventory of therapeutics. Different phylogenetic nomenclatures have been used by different entities (e.g., NextStrain, GISAID, Pangolin, Public Health England, and WHO) for the circulating SARS-CoV-2 strains. The Spike genes have undergone many missense mutations and deletions, the most dangerous for immune escape being the ones within the ACE2 receptor-binding domain (RBD) (such as K417N/T, N439K, L452R, T478K, E484K/Q, and N501Y) and the furin-cleavage site (such as P681H/R). Convergent evolution has led to overlapping combinations of mutations in distant clades. In this book, we focus on the molecular mechanisms of convergent evolution and summarize *in vitro* and *in vivo* evidences of efficacy for convalescent plasma, currently approved vaccines, and monoclonal antibodies against SARS-CoV-2 variants of concern (VOC: Alpha, Beta, Gamma, Delta), variants of interest (VOI: Lambda and Mu) and other strains under monitoring (Eta, Theta, Iota, Zeta, Kappa, Epsilon).

Contents

Abbreviations

CCP	COVID-19 convalescent plasma
MR	Mutation rate
nAb	Neutralizing antibodies
RBD	Receptor-binding domain
RBM	Receptor-binding motif
RCT	Randomized controlled trials
VOC	Variant of concern
VOI	Variant of interest
VUI	Variant under investigation

Chapter 1
Why the Spike Protein is Relevant for COVID-19 Therapeutics

Abstract The Spike protein critical residues focus on the receptor-binding domain, and more specifically on the receptor binding motif.

Keywords Spike · RBD · RBM

The COVID-19 pandemic driven by SARS-CoV-2 has totaled more than 200 million cases and 4 million deaths worldwide since December 2020 to September 2021. Many prophylactic and therapeutic regimens [1, 2] have been tested in randomized controlled trials (RCT), but to date, only dexamethasone [3] and remdesivir [4] have shown conclusive evidences of clinical benefit.

The Spike (S) protein drives SARS-CoV-2 infectivity: it is a type I fusion glycoprotein, responsible for initiating the infection leading to COVID-19. Approximately 35% of the SARS-CoV-2 S glycoprotein consists of carbohydrate [5]. As a feature unique of SARS-CoV-2, the thick glycan shield covering the S protein is not only essential for hiding the virus from immune detection, but it also plays multiple functional roles, stabilizing the S prefusion open conformation, which is competent for binding the ACE2 primary receptor, and gating the open-to-close transitions [6].

ACE2 variants that enhance and reduce Spike binding have been reported, including two variants with distinct population distributions that enhanced affinity for Spike. ACE2 p.Ser19Pro ($\Delta\Delta G = 0.59 \pm 0.08$ kcal mol^{-1}) is often seen in the African cohort (AF = 0.003), while p.Lys26Arg ($\Delta\Delta G = 0.26 \pm 0.09$ kcal mol-1) is predominant in the Ashkenazi Jewish (AF = 0.01) and European non-Finnish (AF = 0.006) cohorts. Carriers of these alleles may be more susceptible to infection or severe disease, and these variants may influence the global epidemiology of COVID-19. Three rare ACE2 variants strongly inhibited (p.Glu37Lys, $\Delta\Delta G = -1.33 \pm 0.15$ kcal mol^{-1} and p.Gly352Val, predicted $\Delta\Delta G = -1.17$ kcal mol^{-1}) or abolished (p.Asp355Asn) Spike binding and may confer resistance to infection [7].

Each virion harbors 30–40 Spike homotrimers on the envelope [8, 9], with each monomer consisting of two domains (S1 and S2). TMPRSS2, factor Xa and thrombin are recognized to be important for cleavage activation of SARS-CoV-2 Spike [10]. S1 domain includes the receptor-binding domain (RBD), which incorporates the

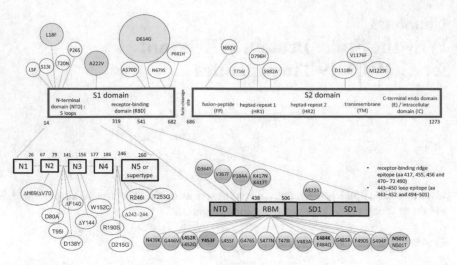

Fig. 1.1 Linearized representation of substitutions and deletions commonly detected in Spike protein

receptor-binding motif (RBM) (Fig. 1.1). Anti-Spike antibodies can be grouped in 11 clusters according to epitopes or in four classes according to mechanism of action (Table 1.1). Passive immunotherapies based on anti-Spike neutralizing antibodies (nAb), which develop in close to 90% of patients and persist for at least 5 months [11], have led therapeutics development. nAbs isolated from convalescents preferentially use specific heavy-chain germline genes, and the two most frequently elicited anti-body families (IGHV3-53/3–66 and IGHV1-2) bind the RBD in two different modes [12]. The first nAb-based manufactured therapeutic has been COVID-19 convales-cent plasma (CCP), whose efficacy seems promising [13, 14] but for which RCTs were not conclusive [15]. Antiviral monoclonal nAbs entered the market at the begin-ning of 2021 [16], and polyclonal IgG formulations (i.e., hyperimmune serum) will likely follow [17]. All such nAb-based therapeutics and vaccines share a common drawback: the risk for selective pressure and mutational escape of the Spike protein [18]. Some of those iatrogenic changes in Spike protein might increase strain trans-missibility, increase re-infection rates or reduce the efficiency of vaccine campaigns [19].

Like many other viral surface proteins, the trimeric SARS-CoV-2 Spike (S) protein is heavily glycosylated with 22 N- and 2 O-glycosites per monomer which are likely to influence S protein folding and evade host immune response. The S protein glycosites are highly conserved, and the glycosites at positions 801 and 1194 are essential for viral entry. In addition, the RBD of S1 and the heptad repeat (HR) regions of S2 contain most of highly conserved sequences [20].

Table 1.1 Competition clusters for anti-SARS-CoV-2 Spike monoclonal antibodies. Modified with permission from references [481–483]

Cluster (adapted from Ref. [117])	Representative mAbs	Class (adapted from Ref. [144])	Epitope (adapted from Ref. [483])	Representative mAbs	Resistant mutation(s)
1	COVA2-16, COVA2-31, COVA2-23, COVA2-11, COVA3-06, COVA3-09, COVA2-29, COVA2-45, COVA1-18, COVA2-20, COVA2-39, COVA2-15	1 (completely block ACE2, accessibility to RBD epitope only in "up" conformation as for ACE2; VH3-53 or VH3-66 with short (<15 aa) CDRH3); strain-specific	**RBM class I**	C102	?
				C105	?
				B38	?
				BD-236	?
				BD-604	?
				BD-629	?
				CV30	?
				CC12.1	?
				CC12.3	?
				COR-101	?
				Bamlanivimab/LY-CoV555	K417N, L452R, E484K, S494P, Q493R
				Etesevimab/LyCoV016/CB6	K417N/T, N460T, A475V, N487K/D
				C07-250	?
				P2C-1F11	?
				S2E12	?
				S2H14	?

(continued)

Table 1.1 (continued)

Cluster (adapted from Ref. [117])	Representative mAbs	Class (adapted from Ref. [144])	Epitope (adapted from Ref. [483])	Representative mAbs	Resistant mutation(s)
				REGN10933	E406W, K417E/V/N, Y453F, L455F, E484K, F486V, Q493K/R
III	COVA2-04, COVA2-13, COVA2-07, COVA2-24, COVA2-44, COVA1-16	**2** (block ACE2, accessibility to RBD epitope in either "up" or "down" conformations; various VH with long (>15 aa) CDRH3); strain-specific	**RBM class II**	C002	?
				C104	?
				C119	?
				C121	?
				COVA2-39	?
				5A6	?
				P2B-2F6	L452R, V483A, F490L
				P2C-1A3	?
				S2H13	?
				H11-H4	?
				H11-D4	?
				Ty1	?
				Sb-23	?
				CV-07-270	?
				BD-368-2	?

(continued)

Table 1.1 (continued)

Cluster (adapted from Ref. [117])	Representative mAbs	Class (adapted from Ref. [144])	Epitope (adapted from Ref. [483])	Representative mAbs	Resistant mutation(s)
			RBM class III (make contact with nearby RBDs, eventually locking the Spike homotrimer)	S2M11	?
				Nb66	?
				Nb20	?
				C144	E484K, Q493R
				Ab2-4	?
				BD23	?
VI	COVA1-01, COVA1-02, COVA1-27, COVA2-34, COVA1-12	3 (does not overlap with ACE2-binding site; accessibility to RBD epitope in "up"/"down" conformation, locking the Spike homotrimer in closed conformation; strain-specific	RBD core cluster I	C135	R346S, N440K
				S309 (cross-neutralizes SARS-CoV)	N501Y
				C110	?
				2-43	?
				REGN10987	E406W N439K, N440K, K444Q, V445A, N450D, L452R
VII	COVA2-02, COVA2-46, COVA2-05	4 (does not overlap with ACE2-binding site; accessibility to RBD epitope only in "up" conformation: promote S1 shedding; various IgHV usage; strain-specific	RBD core cluster II	CR3022 (cross-neutralizes SARS-CoV)	?
				COVI1-6	?
				EY6A (cross-neutralizes SARS-CoV)	?

(continued)

Table 1.1 (continued)

Cluster (adapted from Ref. [117])	Representative mAbs	Class (adapted from Ref. [144])	Epitope (adapted from Ref. [483])	Representative mAbs	Resistant mutation(s)
				S304 (cross-neutralizes SARS-CoV)	?
				VHH72 (cross-neutralizes SARS-CoV)	?
				S2X35	?
				H014 (cross-neutralizes SARS-CoV)	?
				S2A4	?
IX (NTD)	COVA2-25, COVA2-03, COVA2-22, COVA2-30, COVA1-06, COVA2-17, COVA3-07, COVA1-20, COVA2-06, COVA3-05, COVA1-09, COVA2-37, COVA1-22	Strain-specific; interfere with fusion or cause steric hindrance	NTD	5–24	Δ144, Δ242-244, R246I
				4–8	Δ144, Δ242-244, R246I
				4A8	Δ144, Δ242-244, R246I
				2–17	Δ144, Δ242-244
				4–19	Δ144, Δ242-244, R246I

(continued)

Table 1.1 (continued)

Cluster (adapted from Ref. [117])	Representative mAbs	Class (adapted from Ref. [144])	Epitope (adapted from Ref. [483])	Representative mAbs	Resistant mutation (s)
IV	COVA2-40, COVA1-25	Non-classified	AZD8895/COV2-2196 CV07-287		
X	COVA1-03				
XI	COVA1-21				
II	many				
V					
VIII					

Chapter 2
Whole Genome Mutation Rates

Abstract Mutation rates for the entire SARS-CoV-2 genome are lower than for other pathogenic viruses, and will be affected by mass vaccination.

Keyword Mutation rate

Coronaviruses belong to the order Nidovirales, which is known for viruses with the longest RNA genome [21]. SARS-CoV-2 genome includes 29,903 ribonucleotides, which in turn encode 29 proteins. Coronaviruses have a proofreading apparatus [22], but their genomes nevertheless remain prone to recombination as well as other copy-choice transcriptional errors [23]. Being relatively recent, the observed genome diversity is lower than for other RNA viruses [24]. Most proteins exhibit little mutational variability, the proteins with the highest mutation rate (MR) currently being Spike, NSP12 [RNA-dependent RNA polymerase (RdRp)] and NSP9c [25]. The average MR of the entire genome has been estimated from the related mouse hepatitis virus (MHV) at 10^{-6} nucleotides per cycle, or 4.83×10^{-4} subs/site/year, which is similar, or slightly lower, than the one observed for other RNA viruses [26]. Heterogeneous mutation patterns reflect host antiviral mechanisms that are achieved through apolipoprotein B mRNA editing catalytic polypeptide-like proteins (APOBEC), adenosine deaminase acting on RNA proteins (ADAR), ZAP proteins and probable adaptation against reactive oxygen species (ROS) [27]. G $\rightarrow$ U and C $\rightarrow$ U mutations, a well-known result of APOBEC and ROS, are prevalent and occur many times at the same genome positions along the global SARS-CoV-2 phylogeny (a phenomenon also known as homoplasy) [28].

While the specific Spike mutation rates will be discussed in the paragraph below, it is noteworthy that genes other than S mutate in SARS-CoV-2. The global incidence frequency of N:203K/204R has rose up from nearly zero to 76% to date with a shrinking from August to November in 2020 but bounced later. The emergence of B.1.1.7 is associated with the second growth of R203K/G204R mutants. The 203K/204R virus increased the infectivity in a human lung cell line and induced an enhanced damage to blood vessel of infected hamsters' lungs [29].

© The Author(s), under exclusive license to Springer Nature Switzerland AG 2021
D. Focosi, *SARS-CoV-2 Spike Protein Convergent Evolution*,
SpringerBriefs in Microbiology,
https://doi.org/10.1007/978-3-030-87324-0_2

The diversity of the SARS-CoV-2 lineages is declining at the country level with increased rate of mass vaccination ($r = -0.72$). Given that the COVID-19 vaccines leverage B cell and T cell epitopes, analysis of MR shows neutralizing B cell epitopes to be particularly more mutated than comparable amino acid clusters (4.3-fold). Vaccine breakthrough patients harbor viruses with 2.3-fold lower diversity in known B cell epitopes compared to unvaccinated COVID-19 patients. Vaccinated break-through patients also displayed fewer COVID-associated complications and preex-isting conditions relative to unvaccinated COVID-19 patients. COVID-19 vaccines are fundamentally restricting the evolutionary and antigenic escape pathways acces-sible to SARS-CoV-2: the societal benefit of mass vaccination may consequently go far beyond the widely reported mitigation of SARS-CoV-2 infection risk and amelio-ration of community transmission, to include stemming of rampant viral evolution [30].

Chapter 3
Phylogenetic Systems

Abstract Different nomenclatures have been used, replaced by newer schemes as soon as the former got too complicated. Reconciliation is mandatory to interpret early literature.

Keywords Lineages · Clades · PANGOLIN · NextStrain · Public Health England · GISAID · Variants of concern · Variants of interest · Variants under monitoring

In general, the nomenclature accounting for genetic diversity within a given species is not regulated by the International Committee on Taxonomy of Viruses (ICTV). Historically, such low-level genetic diversity has been variably grouped into "clades", "subtypes", "genotypes", "groups" or "lineages". The main repositories for SARS-CoV-2 genomic sequences are listed in Table 3.1.

On April 2020, a preliminary work by the London School of Hygiene and Tropical Medicine on 5300 SARS-CoV-2 sequences from 62 countries identified two clusters (C1 and C2), further stratified in six main clades (**C1**, **C.1.1**, **C2**, **C2.1**, **C2.1.1** and **C.2.1.2**) [31]. Such findings were soon replicated by a Chinese study in June 2020 using only 103 isolates, which first introduced the **L** and **S** lineage nomenclature [32].

The **Global Initiative on Sharing All Influenza Data (GISAID)** repository currently includes more than 2.2 million full SARS-CoV-2 genome sequences (most countries having sequences and shared less than 5% of reported cases, Australia, UK and Denmark representing brilliant exceptions) and classifies clades with progressive letters (https://www.gisaid.org/index.php?id=208). In Winter 2020, the main clades were **L**, **O**, **V** and **S**. Later, clade **G** (with the associated D614G mutation in the Spike protein) emerged followed by the related **GR** and **GH** clades [33]. An eight clade named **GV** has since been described in the following months.

The **Phylogenetic Assignment of Named Global Outbreak LINeages (PANGOLIN)** lineage nomenclature [34, 35] (https://github.com/nextstrain/ncov/blob/master/docs/naming_clades.md) is one of the two main nomenclature systems. Pango lineage names comprise an alphabetical prefix and a numerical suffix. The alphabetical prefix contains Latin characters only which are case insensitive. Each

Table 3.1 Main SARS-CoV-2 gene sequence repositories and analysis tools

Repository	URL
China National Center for Bioinformation (CNCB)—National Genomics Data Center (NGDC)	https://bigd.big.ac.cn/ncov/release_genome?lang=en
China National Microbiology Data Center (NMDC)	http://nmdc.cn/nCov/en
COVID-19 genomics consortium UK (CoG-UK)	https://www.cogconsortium.uk/
Global initiative on sharing all influenza data (GISAID)	https://www.gisaid.org/epiflu-applications/phylodynamics/
NCBI SARS-CoV-2 GenBank	https://www.ncbi.nlm.nih.gov/labs/virus/vssi/#/virus?SeqType_s=Nucleotide&VirusLineage_ss=SARS-CoV-2,%20taxid:2697049
NextStrain [36]	https://nextstrain.org/sars-cov-2
NextStrain on JBrowse 2	https://nextstrain-jbrowse.herokuapp.com/ncov/global
Analysis tools	
Virus pathogen resource (ViPR)	https://www.viprbrc.org/brc/vipr_genome_search.spg
SARS-CoV-2 mutation browser v-1.3 [484]	http://covid-19.dnageography.com/
Microbial genome mutation tracker (MicroGMT) [485]	https://github.com/qunfengdong/MicroGMT
Coronapp [486]	http://giorgilab.unibo.it/coronannotator/
Ensembl Variant Effect Predictor [487]	https://www.ensembl.org/info/docs/tools/vep/index.html
PrimerChecked	https://epicov.org/epi3/frontend#11daab (registration required)
Infection pathogen detector 2.0 [488]	http://ipd.actrec.gov.in/ipdweb
Coronavirus antiviral research database (CoV-RDB) [489]	https://covdb.stanford.edu/
Lineage assignment	
NextClade	https://clades.nextstrain.org/
Pangolin COVID-19 Lineage assigner	https://pangolin.cog-uk.io/

(continued)

Table 3.1 (continued)

Repository	URL
US SARS-CoV-2 variant dashboard [490]	https://janieslab.github.io/sars-cov-2.html
CovRadar [491]	https://gitlab.com/dacs-hpi/covradar
Rapid assessment of SARS-CoV-2 Clades (RASCL) [492]	https://github.com/veg/SARS-CoV-2_Clades
Powered by GISAID	
CoV spectrum	https://cov-spectrum.ethz.ch/
Global evaluation of SARS-CoV-2/hCoV-19 sequences (GESS) [493]	https://wan-bioinfo.shinyapps.io/GESS/
Regeneron COVID-19 dashboard	https://covid19dashboard.regeneron.com/
CoVariants	https://covariants.org/
Covid-miner	https://covid-miner.ifo.gov.it/
CoVizu [494]	http://filogeneti.ca/covizu/
COVID-19 CoV genetic browser	https://covidcg.org/?tab=location
NAAT amplicons	https://covid-19-diagnostics.jrc.ec.europa.eu/amplicons
COVID-19 viral genome analysis pipeline	https://cov.lanl.gov/content/index
CoV-GLUE	http://cov-glue.cvr.gla.ac.uk/#/home
Genomic signature analysis	https://covid19genomes.csiro.au/index.html#
Geographic mutation tracker	https://www.cbrc.kaust.edu.sa/covmt/
Interactive real-time mutation tracker	https://users.math.msu.edu/users/weig/SARS-CoV-2_Mutation_Tracker.html
Global testing and genomic variability	https://bioinfo.lau.edu.lb/gkhazen/covid19/genomics.html
Spike protein mutations monitoring	https://www.molnac.unisa.it/BioTools/cov2smt/index.php
ViruSurf	http://geco.deib.polimi.it/virusurf_gisaid/
Outbreak.info [495]	https://outbreak.info

(continued)

Table 3.1 (continued)

Repository	URL
SpikePro [496]	http://github.com/3BioCompBio/SpikeProSARS-CoV-2

Modified with permission from reference [481]

dot in the numerical suffix means "descendent of" and is applied when one ancestor can be clearly identified. So, lineage B.1.1.7 is the seventh named descendent of lineage B.1.1, and C.1 is the first named descendent of lineage C. The suffix can contain a maximum of three hierarchical levels, referred to as the primary, secondary and tertiary suffixes. In order to avoid four or more suffix levels, a new lineage suffix is introduced, which acts as an alias. All top-level lineages that are recombinants have a prefix that begins with X.

NextStrain [36] sources data from public repositories (such as GISAID, NCBI, ViPR and GitHub) and supports the year-letter dynamic PANGOLIN naming code. Clades originally needed a frequency of at least 20% globally for two or more months and are named with the year it was first identified and the first available letter within the alphabet. The parent clade is reported with the "." notation (e.g., 19A.20A.20C to indicate clade 20C). Then, in January 2021, it was acknowledged that lack of international travel made it slower for new clades to move past 20% global frequency, and consequently, two alternative requirements were added: clade reaches >20% global frequency for two or more months: a clade reaches >30% regional frequency for two or more months, and a **VOC ("variant of concern")** is recognized [37].

A distinct nomenclature for variants has been implemented by **Public Health England (PHE)**, represented by last two digits of the year, first three capital letters of the month and progressive (two digits) number of variants reported in such month–year (e.g., "VOC-21FEB-02").

Finally, at the end of May 2021, considering that the abovementioned systems are difficult to say and recall and are prone to misreporting, and this had induced people to call variants by the place where they had been detected (which could be stigmatizing and discriminatory for countries that should instead be plauded for their higher sequencing efforts), the WHO issued a simplified nomenclature based on the Greek alphabet. Additionally, WHO [moving from the former NextStrain and PHE definitions of VOC and **variants under investigation (VUI)**], defined a SARS-CoV-2 isolate a **variant of interest (VOI)** if, compared to a reference isolate, its genome has *"mutations with established or suspected phenotypic implications, either has been identified to cause community transmission/multiple COVID-19 cases/clusters, or has been detected in multiple countries OR is otherwise assessed to be a VOI by WHO in consultation with the WHO SARS-CoV-2 Virus Evolution Working Group"*. A VOC meets the definition of a VOI but is also *"associated with increase transmissibility or detrimental change in COVID19 epidemiology or increase in virulence or change in clinical disease presentation or decrease in effectiveness of public health and social measures or available diagnostics vaccine, or therapeutics"*. VOC and VOI. A previously designated VOI or VOC which has conclusively demonstrated to no longer pose a major added risk to global public health compared to other circulating SARS-CoV-2 variants can be reclassified. Alerts for further monitoring are defined as variants with genetic changes that are suspected to affect virus characteristics with some indication that it may pose a future risk, but evidence of phenotypic or epidemiological impact is currently unclear, requiring enhanced monitoring and repeat assessment pending new evidence. (https://www.who.int/en/activities/tracking/SARS-CoV-2-variants/). The US CDC, when accepting such classification, defined the most disruptive VOCs as *"variant of high consequence"* (https://www.cdc.gov/coronavirus/2019-ncov/var iants/variant-info.html), while the European CDC additionally classified the less concerning WHO VOI and many more strains as **"variants under monitoring (VUM)"**, *"detected as signals through epidemic intelligence, rules-based genomic variant screening, or preliminary scientific evidence"* (https://www.ecdc.europa.eu/ en/covid-19/variants-concern).

The different SARS-CoV-2 phylogenies are reconciled in Table 3.2, which details the separating (barcoding) SNPs. Globally, the year 2020 marked a positive selection of D614G, S477N (clade 20A.EU2), A222V (20A.EU1) and V1176F SNPs, an expansion of B.1 clade, especially strain containing Q57H (B.1.X), N:R203K/G204R (B.1.1.X), T85I (B.1.2–B.1.3), G15S+T428I (C.X) and I120F (D.X) [38]. None of the SARS-CoV-2 variants described so far has definitively been shown to increase infection severity; on the contrary, a clade 19B variant with lower severity was detected in Singapore in the Spring 2020 and then disappeared [39].

Viruses with both S:D614G and RdRp:P323L mutations have lower ratios of non-synonymous mutations per non-synonymous site to synonymous mutations per synonymous site (dN/dS) compared to those without the two mutations, particularly at RdRp coding region and Orf8 gene. Instead, S gene had higher dN/dS ratios in the mutant genomes. While the S gene was under stronger negative selection in wild-type genomes during the early stages, it is almost at equal levels between mutant and wild-type genomes in the later stages. Instead, RdRp is under stronger overall negative selection in the mutant genomes, particularly during the early stages [40]. It has been estimated that, as of March 2021, the current SARS-CoV-2 variants of concern (VOC, see below) have sampled only 36% of the possible Spikes changes which have occurred historically in Sarbecovirus evolution [41].

Table 3.2 Main clades/lineages according to different naming schemes

London clade	GISAID clade	PANGOLIN lineage	NextStrain Clade	Originary country	Separating (barcoding) non-synonymous single nucleotide mutations and deletions	Corresponding effects on protein sequence	Max frequency
C1	L	**B.3–B.7, B.9, B.10, B.13–B.16**	19A	Asia: China/Thailand	**Root clade**		65-47% globally in January 2020, now disappearing
n.a.	O				Others		
n.a.	V	B.2			G11083T	NSP6:L37F	Now disappearing
C1.1	n.a.	n.a.	n.a.		C18060T A17858G	orf1ab:nsp14:S7F orf1ab:nsp13:M541V	?
C2	S	A.1–A.6	19B	Asia: China	C8782T	NSP4:S76S (synonymous)	Contains "reference sequence" WIV04/2019;
					T28144C	ORF8:L84S	
					G26144T	ORF3a:G251V	28-33% globally in January 2020; now in some restricted areas in the USA and Spain, but resurging in February 2021 thanks for convergent evolution [497, 498]

(continued)

Table 3.2 (continued)

London clade	GISAID clade	PANGOLIN lineage	NextStrain Clade	Originary country	Separating (barcoding) non-synonymous single nucleotide mutations and deletions	Corresponding effects on protein sequence	Max frequency
n.a.	G/478K.V1	**B.1.617.2** **AY.1** **AY.2**	21A	India	**See text**	See text	Since October 2020
C.2.1	G	**B.1.5–B.1.72**	20A	N America/Europe/Asia: USA, Belgium, India	**C14408T** **A23403G** (C241T) (C3037T)	NSP12b:P314L S:D614G 5'UTR NSP3:F106F	Found in Germany, Australia and China in January 2020; basal pandemic lineage bearing S 614G that's globally distributed
	GH	**B.1.9, B.1.13, B.1.22, B.1.26, B.1.37**		?	?	?	?
C.2.1.1		**B.1.2–B.1.66**	20C (US)	N America: USA (includes CAL.20C and B.1.525)	C14408T A23403G (C241T) (C3037T) **G25563T** **C1059T**	NSP12b:P314L S:D614G 5'UTR NSP3:F106F ORF3a:Q57H orf1ab:nsp:T85I	Derived from 20A since February 2020; southern U.S. in late May of 2020 [499]; globally distributed

(continued)

Table 3.2   (continued)

London clade	GISAID clade	PANGOLIN lineage	NextStrain Clade	Originary country	Separating (barcoding) non-synonymous single nucleotide mutations and deletions	Corresponding effects on protein sequence	Max frequency
n.a.			20G	USA	Many	ORF1b 1653D ORF3a 172V N 67S N 199	Derived from 20C, main strain in USA in second wave
n.a.		B.1.351	20H (V2) (formerly 20H/501Y.V2)	South Africa	Many	D80A D215G K417N E484K N501Y A701V	Derived from 20C, concentrated in South Africa
n.a.		B.1.466.2	n.a	Indonesia	See text	See text	Since November 2020
n.a.		B.1.526	21F	USA	See text	See text	Since November 2020
n.a.		B.1.621	21H	Colombia	See text	See text	Since January 2021
C2.1.2	GR	B.1.1	20B	Europe: UK, Belgium, Sweden (includes B.1.207)	C14408T A23403G (C241T) (C3037T) **GGG28881AAC**	NSP12b:P314L S:D614G 5′UTR NSP3:F106F N:RG203KR	Derived from 20A since February 2020; Globally distributed

(continued)

Table 3.2 (continued)

London clade	GISAID clade	PANGOLIN lineage	NextStrain Clade	Originary country	Separating (barcoding) non-synonymous single nucleotide mutations and deletions	Corresponding effects on protein sequence	Max frequency
n.a.			20D		Many	ORF1a 1246I ORF1a 3278S	Derived from 20B, concentrated in South America, southern Europe and South Africa
n.a.		P.1 P.2	20J (V3)	Brazil	See text	See text	
n.a.		n.a.	20F	n.a.	Many	ORF1a 300F S 477N	Derived from 20B, concentrated in Australia
n.a.	GR/452Q.V1	C.37	21G	Peru	See text	See text	Since December 2020

(continued)

Table 3.2 (continued)

London clade	GISAID clade	PANGOLIN lineage	NextStrain Clade	Originary country	Separating (barcoding) non-synonymous single nucleotide mutations and deletions	Corresponding effects on protein sequence	Max frequency
n.a.	**GRY**		**20I (V1)**	South-East UK (includes **B.1.1.7/20I/501Y.V1**	ORF1ab:C3267T ORF1ab:A1708D ORF1ab:I2230T ORF1ab: Δ11288-11296 S:21765-21770 deletion S:21991-21993 deletion S:A23063T S:C23271A S:C23604A S:C23709T S:T24506G S:G24914C Orf8:C27972T Orf8:G28048T Orf8:A28111G N:28280 GAT->CTA N:C28977T	ORF1ab:T1001I ORF1ab:A1708D ORF1ab:I2230T ORF1ab: ΔSGF3675-3677 S: ΔHV69-70 S: ΔY144 S:N501Y S:A570D S:P681H S:T716I S:S982A S:D1118H ORF8:Q27stop ORF8:R52I ORF8:Y73C N:D3L N: S235F	10% in UK December 2020 derived from 20B
n.a.	**GV**	**B.1.177**	**20E (EU1)** (formerly 20A/EU.1)	Spain	Many	N 220V ORF10 30L ORF14 67F A222V many	Main strain in second wave in EU Derived from 20A

n.a. Not available. Modified with permission from reference [481]

Chapter 4
Mechanism of Immune Escape: Single Nucleotide Mutations, Insertion/Deletions and Recombination

Abstract Single nucleotide mutations as well as insertion and deletions occur randomly in the genome. Super-infection (co-infection) predisposes to recombination.

Keywords SNP · Insertion · Deletion · Co-infection · Super-infection · Recombination

The main forces of viral evolution, i.e., mutation, drift, recombination and selection, all operate within hosts [42], and the global SARS-CoV-2 population is actually a meta-population consisting of the viruses in all the infected hosts. Advantageous mutations can occur in individual patients and then expand at a global scale.

Single nucleotide polymorphisms (SNP) can lead to missense mutations (summarized in Table 3.2). As of February 2021, there were 2592 distinct SARS-CoV-2 variants [43]. 95% of patients show within-host diversity, mostly due to mutational hot spots [44]. High-confidence subclonal variants were found in about 15.1% of the NGS data sets, which might indicate co-infection from different strains and/or intra-host evolution [43]. SNPs are rare because of proofreading efficiency of the SARS-CoV-2 RNA-dependent RNA polymerase (nsp12) and the error-correcting exonuclease protein non-structural protein 14 (nsp14): P203L mutation in nsp14 almost doubles the genomic MR (from 20 to 36 SNPs/year) [45].

There were as many as 420 unique indel positions and 447 unique combinations of indels. Despite their high frequency, indels resulted in only minimal alteration, including both gain and loss, of N-glycosylation sites. Indels and point mutations are positively correlated, and sequences with indels have significantly more point mutations [46].

Deletions drive sudden antigenic drift and compromise binding of nAb [47]: Deletions in the N-terminal domain (such as ΔH69/ΔV70 and ΔY144) are increasingly prevalent [48]. 90% of deletions maintain the reading frame and fall within four regions (RDRs) within the NTD at positions 60–75 (RDR1), 139–146 (RDR2), 210–212 (RDR3) and 242–248 (RDR4) of the S protein [47]. E.g., SARS-CoV-2 lineages circulating in Brazil with mutations of concern in the RBD independently acquired convergent deletions and insertions in the NTD of the S protein, which altered the NTD antigenic-supersite and other predicted epitopes at this region [49]. The BriSΔ variant, originally identified as a viral subpopulation by passaging SARS-CoV-2 isolate hCoV-19/England/02/2020, has an in-frame 8 amino acid deletion in Spike encompassing the furin recognition motif and S1/S2 cleavage site (amino acids 679–687 NSPRRARSV, replaced by I) [50].

Garushyants et al. identified 141 unique **inserts** of different lengths. These inserts reflect actual virus variance rather than sequencing errors. Two principal mechanisms appear to account for the inserts in the SARS-CoV-2 genomes: polymerase slippage and template switch that might be associated with the synthesis of subgenomic RNAs. Inserts in the Spike glycoprotein can affect its antigenic properties and thus have to be monitored. At least, two inserts in the N-terminal domain of the Spike (ins246DSWG and ins15ATLRI) that were first detected in January 2021 are predicted to lead to escape from nAbs, whereas other inserts might result in escape from T cell immunity [51].

There are putative [44] and in vivo [52] evidences of **super-infection or co-infection** from different SARS-CoV-2 strains. While studies relying on linkage disequilibrium have identified that **recombination** occurs at very low levels [52, 53] or does not occur at all [32, 54–58], a new method detected multiple recombination events using relatively small samples [59]. Nevertheless, recombination rates across the genome of the human seasonal coronaviruses 229E, OC43 and NL63 vary with rates of adaptation [60]. Jackson et al. presented evidence for multiple independent origins of recombinant SARS-CoV-2 sampled from late 2020 and early 2021 in the United Kingdom. Their genomes carried SNP and deletions that were characteristic of the B.1.1.7 VOC, but lacked the full complement of lineage-defining mutations. Instead, the rest of their genomes share contiguous genetic variation with non-B.1.1.7 viruses circulating in the same geographic area at the same time as the recombinants. In four instances, there was evidence for onward transmission of a recombinant-origin virus, including one transmission cluster of 45 sequenced cases over the course of two months. The inferred genomic locations of recombination breakpoints suggest that every community-transmitted recombinant virus inherited its Spike region from a B.1.1.7 parental virus, consistent with a transmission advantage for B.1.1.7's set of mutations [61]. Turakhia et al identified 606 recombination events by investigating a 1.6M sample tree: approximately 2.7% of sequenced SARS-CoV-2 genomes have recombinant ancestry, and recombination breakpoints occur disproportionately in the Spike protein region. The coinfection indices of GISAID20May11 and GISAID21Apr1 (2021) datasets were 16 and 34, respectively: there was a linear relationship between the coinfection index and the coinfected variant numbers, and cases were coinfected with 2.20 and 3.42 SARS-CoV-2 variants on average. A large study identified 53 (~0.18%) co-infection events (including with 2 Delta sublineages) out of 29,993 samples: apart from 52 co-infections with 2 SARS-CoV-2 lineages, one sample with co-infections of 3 SARS-CoV-2 lineages was firstly identified. Another study identified coinfections around 0.61% of all samples investigated (9 cases).

Through RNA-seq, **chimeric host-virus reads** can be detected in the infected cells. But further analysis using mixed libraries of infected cells and uninfected zebrafish embryos demonstrates that these reads are falsely generated during library construction. In support, whole genome sequencing also does not identify the existence of chimeric reads in their corresponding regions. Therefore, the evidence for SARS-CoV-2's integration into host genome is lacking [62].

All the major VOCs harbor the deletion in ORF1ab (del11288-11296 (3675–3677 SGF) [63] and have signals of positive selection in Spike (convergent for 16 sites and non-convergent for five sites) [64]. Given consistent convergent evolution, we will separately discuss individual mutations first and will later focus on variants. It should not be forgotten that among the deleterious variants (impacting mortality), NSP3 had the highest incidence followed by NSP2 (T265I), ORF3a (Q57H), N (R203K and G204R) and finally Spike [65, 66]. Interestingly, the SARS-CoV-2 genome was more susceptible to mutation because of the high frequency of nt14408 mutation (which located in RNA polymerase) and the high expression levels of ADAR and APOBEC in severe clinical outcomes [66].

The SARS-CoV-2 ancestral strain has caused pronounced superspreading events, reflecting a disease characterized by overdispersion, where about 10% of infected people causes 80% of infections. Lockdowns exert an evolutionary pressure which favors variants with lower levels of overdispersion. Overdispersion is an evolutionarily unstable trait, with a tendency for more homogeneously spreading variants to eventually dominate [67].

Chapter 5
Spike Protein Mutations Detected in Currently Circulating Strains

Abstract The main mutations in Spike are discussed individually in much detail, moving from structure to function.

Keywords Mutation rate · RBD · RBM · FCS

Structurally, the Spike protein of SARS-CoV-2 has an additional cleave in the S1 subunit compared to SARS-CoV-1. A first study reported the nucleotide mutation rate (MR) of Spike gene from January to April 2020 at 2.19×10^{-3} substitution/site/year [68], which was significantly higher than the MR of the entire genome [69, 70]. At 9-months, such MR remained unvaried at 1.08×10^{-3} ribonucleotide substitutions/site/year, similar across clades [71].

Borges et al estimated SARS-CoV-2 mutation rate and demonstrate the repeatability of its evolution when facing a new cell type but no immune or drug pressures at 3.7×10^{-6} nt^{-1} $cycle^{-1}$ for a lineage of SARS-CoV-2 with the originally described Spike protein and of 2.9×10^{-6} nt^{-1} cycle-1 for a lineage carrying the D614G mutation that has spread worldwide. Mutation accumulation is heterogeneous along the genome, with the Spike gene accumulating mutations at a mean rate 16×10^{-6} nt^{-1} per infection cycle across backgrounds, 5-fold higher than the genomic average. Mutators emerged in the D614G background, likely linked to mutations in the RNA-dependent RNA polymerase and/or in the proofreading exonuclease [72].

The global frequencies of different immune escape mutations have been assessed in several research articles [73]. It has been hypothesized that Spike protein mutations in novel SARS-CoV-2 "variants of concern" commonly occur in or near indels [74]. All 22 N-glycan sites of SARS-CoV-2 Spike remain highly conserved among the variants B.1.1.7, P.1 and 501Y.V2, opening an avenue for robust therapeutic intervention [75]. Increased virulence is more likely to be due to the improved stability to the S trimer in the opened state (the one in which the virus can interact with the cellular receptor ACE2) than due to alterations in the complexation RBD-ACE2 [76]. The consequences of mutations can be dramatic: e.g., high-frequency Spike mutations R346K/S, N439K, G446V, L455F, V483F/A, E484Q/V/A/G/D, F486L,

F490L/V/S, Q493L and S494P/L might compromise some of mAbs in clinical trials [2]. In this paragraph, we will review notable missense mutations and deletions.

D614G appeared in January 2020 and showed a MR of 0.999 in October–November 2020 [25], meaning it fastly became almost universal. D614G compromises the hydrogen bond with T859 of the adjacent monomer, provides higher flexibility, potentially modifies glycosylation at N616 [77], changes the inner motion of the RBD modifying its cross-connections with other domains [78], affects the pH-dependent responsiveness of SARS-CoV-2 and enhances its lysosomal trafficking [79]. Clade G, which derives its name from D614G, and its related strains GR and GH, are characterized by reduced S1 shedding, higher replication in nasopharynx and trachea [80] and increased infectivity [81]: It increases syncytium formation and viral transmission via enhanced furin-mediated Spike cleavage [82]. D614G has lower association with the proteins UGGT1, calnexin, HSP7A and GRP78/BiP (which ensure glycosylation and folding of proteins in the ER), is not cleaved endoproteolytically, binds less laminin [83] and eliminates the unusual cold-induced unfolding characteristics [84]. D614 is nevertheless not worrying for nAb-based therapeutics and vaccines, and it actually increases the susceptibility to neutralization [85, 86]. D614G first established in countries where transmission rates at the beginning of the pandemic were higher [31, 87] and massively expanded. The P323L mutation in the RNA-dependent RNA polymerase (often referred as NSP12b:P314L) accompanies the D614G Spike mutation in most of the analyzed sequences (MR=0.994) [25].

The mutations A222V and L18F are far from the main D614G mutation and are found in the N-terminal domain of the S1 subunit, within areas defined as possible B cell epitopes [88]. The **A222V** mutation (which characterizes the 20A.EU1 clade [89]) was already detected in March 2020 in Iran, expanded in Spain from June to August 2020 (MR from 0.42 to 0.87) and continued its expansion to Norway (MR=0.40), Italy (MR=0.27), Latvia (MR=0.24), Switzerland (MR=0.22), the United Kingdom (MR=0.18) and other European countries. The sequences in October–November yielded MR values of ~0.66-0.72) [25].

L18F in the Spike was marginally present in different countries in March 2020 (MRs ~0.005) [25], but massively spread within the B.1.1.7, P.1 and B.1.351 VOCs detailed below [90].

S31F and **S50L** are located in the NTD. As of March 29, 2021, they have been seen to evolve together in lineage B.1.5.96 in the USA and B.1.1.70.1 (AP.1) in Saxony and moreover have occurred independently before [91]. At the same time, S31F has been detected in 17 viruses in several countries and lineages, likely to have evolved convergently. S50L has been detected in 123 viruses in several countries and lineages.

Δ69-70HV was originally described in association with B.1.1.7. Convergent evolution started in Thailand and Germany in January 2020. ΔH69/ΔV70 diminishes protrusion of the 69–76 loop, increasing Spike-mediated infectivity by 2 folds. Interestingly for screening purposes, the deletion causes false negativity in the Spike target (so-called **S-dropout variant or S-gene target failures (SGTF)**) of a 3-target TaqPath® RT-PCR COVID-19 assay (Thermo Fischer Scientific) [92–94]. The deletion can also be detected as a positive signal using a pair of molecular beacons paired with loop mediated isothermal amplification (LAMP) [95]. B.1.1.7 is associated with higher viral loads [96]. The deletion rarely occurs as a lone mutation [48]. It was later detected in lineages B.1.375 [97, 98] and B.1.346 reported from USA [97] and in lineages B.1.1.7 (described below), B.1.1.298 (described below), B.1.177 (EU1), B.1.160 (EU2) and B.1.258Δ [99] reported from Europe. The deletion causes partial resistance to neutralization by the COVA1-21 mAb, but <3-fold reduction in neutralization by former convalescent sera [100].

T95I has been reported in VOI B.1.526 and B.1.617.1 v.1, in AV.1 and B.1.1.318 and occasionally in the VOC B.1.617.2.

G142D is found in B.1.617 lineage from India [101].

ΔY144 is found in B.1.1.318, A.VOI.V2, B.1.616 and B.1.620.

W152C is found in B.1.427/B.1.429 lineages from USA and many more. That NTD is a region exhibiting particularly high frequency of mutation recruitments, suggesting an evolutionary path on which the virus maintains optimal efficiency of ACE2 binding combined with the flexibility facilitating the immune escape [102].

R246I occurs in B.1.351 v.1

A570D found in B.1.1.7 introduced a salt bridge switch that could modulate the opening and closing of the RBD [103].

Q677H has been observed in eight distinct lineages (including B.1.429 and B.1.525) and is increasing in prevalence [89, 104].

N679S has been found in a few isolates in the US mid-Atlantic region [105].

A suboptimal furin cleavage site (FCS) is located at the $S1/S2$ junction within the sequence $_{681}$PRRAR/S$_{686}$. Such site is not found in related coronaviruses. It promotes infection of respiratory epithelial cells and transmission in animal models [106-108]:

- P681 affects one of **P681H** that has been found both in the UK B.1.1.7 and in the Philippines B.1.1.28 lineage described in details below, in B.1.1.207 lineage in Nigeria [109], in a lineage in Israel [110], in 3 lineages (B.1.243/20A, B.1.222/20B and B.1/20C) in New York [111] and in B.1.1.318, B.1.342.1, P.3/B.1.1.28.3/PHL-B.1.1.28, AV.1, A.VOI.V2, B.1.519 and B.1.621. While P681H may increase Spike cleavage by furin-like proteases, this does not significantly impact viral entry or cell-cell spread [112].
- **P681R** has been found in B.1.617 from India [101, 113]: It facilitates the furin-mediated Spike cleavage and enhances and accelerates cell-cell fusion [114]. P681R mutation—which may affect viral infection and transmissibility—needs to occur on the background of other Spike protein changes to enable its functional consequences [115]. Reverting the P681R mutation to wild-type P681 significantly reduced the replication of Delta variant, to a level lower than the Alpha variant harboring P681H.

T716I has been reported within the B.1/20C clade from New York [111].

D796H has been found in B.1.1.318.

F888L is the peculiar mutations of B.1.525 described below.

T1117I occurs in lineage B.1.1.389 circulating in 2020 in Costa Rica [116]

The RBD is the hot spot of neutralization. Despite RBD-binding antibodies comprise a relatively modest proportion of all Spike-binding IgG serum antibodies in naturally infected individuals (consistent with studies reporting that less than half of Spike-reactive B cells and monoclonal antibodies bind to RBD [117–120]), RBD-binding antibodies contribute the majority of the neutralizing activity in most convalescent human sera [121, 122], both at early (~30 day) and late (~100 day) time points post-symptom onset [123]. There are 56 individual amino acid changes between the RBD of SARS-CoV-2 and SARS-CoV [124], including sites at which antibody escape has been observed for SARS-CoV [125], which explains why the majority of SARS-CoV-induced neutralizing mAbs do not to neutralize SARS-CoV-2 and vice versa. Mutations in the RBD (residues 333-527) outside the RBM have been described.

R346S causes resistance to class 3 antibodies.

V367F has no effect [126] or improves ACE2 affinity via enhanced hydrogen-bonding interactions [18, 127] according to different reports. It is found in the A.23.1 lineage (reported from an Uganda prison in July 2020) together with F157L and Q613H [128].

P384A abrogates neutralization by COVA1-16 mAb [100]. **P384L** lies in the RBD. This mutation has been detected in 799 viruses in 35 countries and multiple lineages (including B.1.1.70.1/AP.1 from Saxony [91]), yet is still present in less than 0.5 % of cases worldwide.

R403T significantly reduced ACE2-mediated virus infection, while a single T403R mutation allows the RaTG13 S to utilize the human ACE2 receptor for infection of human cells and intestinal organoids [129].

K417 has two different significant mutations. **K417N** occurs in the B.1.351 and B.1.617.2.1 VOCs, while **K417T** occurs in the P.1 VOC. Both mutations break the hydrogen bond with ACE2 reducing affinity [130]. Despite the loss in the binding affinity (1.48 kcal/mol, i.e., 6.4-fold drop [131]) between RBD and ACE2 [132], the K417N/T mutations abolish a buried interfacial salt bridge between RBD and escapes neutralization by mAbs etesevimab [131, 133, 134], COVA2-07 and the public COVA2-04 [100, 135, 136], but mutations only modestly affect binding by a few CCP samples [137]. K417R leads to resistance to the REGN-COV2 cocktail [138]. Five out of the 17 most potent mRNA vaccine-elicited mAbs were at least ten-fold less effective against pseudotyped viruses carrying the K417N mutation compared to K417strain [139].

For what concerns immune escape, the most dangerous mutations are the ones within the RBM (residues 438-506): ACE2 binding is increased by mutations at L455, A475, F486, Q493 and P499 and reduced by changes at R439, L452, T470, E484, Q498 and N501 [140]. Mutations within the RBM increasing affinity to the ACE2 receptor definitively deserve special attention. Nevertheless, the four RBM mutations that to date have the highest frequency among sequenced viruses (N439K, Y453F, S477N and N501Y) do not strongly affect binding by convalescent sera [123].

N439K has 2-folds higher binding affinity to ACE2, but this does not translate in higher replication kinetics or clinical severity. First identified in lineage B.1 in March 2020 in Scotland, it is now widespread in association with the ΔH69/ΔV70 deletion, e.g., in B.1.258Δ [99]. N439K mutation causes resistance to different class

3 monoclonal nAbs, including imdevimab (REGN10987) [131] [141], as well as from 8% of convalescent sera [141].

N440K has higher affinity to ACE2 [142, 143]. It emerged during replication in the presence of the class 3 mAb C135 and is resistant to both C135 [139, 143] and REGN10987 [139, 142, 144, 145], but retains full sensitivity to both C121 and C144 [143]. N440K is found in B.1.1.420 [146] and has a frequency of 2.1% in India and a high prevalence in the state of Andhra Pradesh (33.8%) and Karnataka [147] as of March 2021: The variant site was homoplastic, and the variant was found in genomes belonging to different clades (e.g., B.1.36 [147]) and haplotypes. Interestingly, N440K co-occurs with C64F mutation in M glycoprotein [148]. N440K was associated with asymptomatic reinfection of two healthcare workers in Northern India [149] and one individual from Andhra Pradesh [150]: It had been reported in 11 countries as of January 2021 [151] (including Latvia [152], USA [152] and Italy). mRNA vaccines sera neutralizing activity is reduced by 2-folds [153]. The N440K variant produced ten times higher infectious viral titers than a prevalent A2a strain and >1000 folds higher titers than a much less prevalent A3i strain prototype in Caco2 and Calu-3 cells. Interestingly, A3i strain showed the highest viral RNA levels, but the lowest infectious titers in the culture supernatants, indicating the absence of correlation between the RNA content and the infectivity of the sample [154].

Several mutations hit the so-called 443–450 loop epitope (aa 443–452 and 494–501):

- **G446V** mutation reduces neutralization by convalescent serum by 30-folds [137] but only <5 folds in another where KVG444-6TST was tested [100]; neutralization by COVA2-29 mAb was very reduced in the latter study [100].
- **L452R** does not have a major impact in ACE2 affinity when tested in the context of recombinant monomeric RBD but presents enhanced binding within the context of full-length membrane anchored Spike [155]. L452R causes resistance to bamlanivimab [133], while the related mutation L452K causes resistance to COVA2-29 mAb [100]. L452R is the only Spike mutation found in CAL.20A [156] (also known as B.1.232, which also infected gorillas in San Diego zoo [157]) and the most concerning and recently acquired mutation in the CAL.20C (B.1.427/B.1.429) strain from Southern California [158]. L452R is also found in A.21, A.2.4, A.2.5, B.1.1.10, B.1.1.130, B.1.617.1 and B.1.617.2 VOCs [101], B.1.362+L452R, C.16 and C.36. Of interest, the C.37 VOI and a single B.1.74 strain harbors the **L452Q** mutation [156]. L452R also causes evasion from HLA-A24-restricted CTL response [159]. L452Q increases ACE2 binding by 3-folds and *in vitro* infectivity by 2-folds [160].
- **S494P** increases the complementarity between the RBD and ACE2 [127]. S494 interacts with nAb but not with ACE2 [161]. S494P causes ~3 to 5-fold decreases in neutralization titer for a few convalescent sera [137, 161] and escape to several mAbs [162]. It has been isolated in 369 B.1.1.7 sequences from UK from November 12, 2020, to February 5, 2021 [90] and in the B.1/20C clade from New York [111]. **S494D** destroys neutralization activity by both COVA2-29 and COVA1-12 mAbs [100].

- **N501Y** (nicknamed "**Nelly**") falls within an epitope defined by the "443–450" loop, introduces a favorable π-π interaction [103] and increases affinity to ACE2 by 10-folds [142, 163–166] (estimate at −0.81 kcal/mol [167] or −20 kcal/mol [168] in different studies) because of higher number of interactions with residues Y41 and K353 (ACE2) [169, 170]. It also reduces presentation across the majority of MHC-II alleles [171]. N501Y allows a potential aromatic ring–ring interaction and an additional hydrogen bond between the RBD and ACE2 [164], leading to 9-fold stronger binding [132]. N501 in SARS-CoV-2 corresponds to S487 in SARS-CoV, one of the residues whose mutations allowed the species jump from palm civet to humans [172]. N501Y was selected in six passages in aged mice [173, 174] and increases transmissibility and virulence in a murine model [175]. First isolated in Brazil and USA in April 2020 [48], N501Y causes resistance to mAb COV2-2499 [137], modest effects on binding by majority of other mAbs [136] (e.g., COVA1-12 and COVA2-17 [100] or bamlanivimab/LY-CoV555 [164]) and minor reductions in neutralization by convalescent sera [100, 137, 174] or sera from individuals vaccinated with BNT162b2 [164]. Four out of the 17 most potent mRNA vaccine-elicited mAbs were at least 10-fold less effective against pseudotyped viruses carrying the N501Y mutation [139], but another study reported that vaccine-elicited sera were able to neutralize a mouse-adapted SARS-CoV-2 N501Y strain [174]. N501Y is among the main mutations of different variants of concern, i.e., B.1.1.7 from UK, B.1.351 from South Africa and P.1 from Brazil (so that NextStrain initially named those variants as 501Y.Vx), and of B.1.1.70.1/AP.2 from Saxony [91]. On December 2020, a different mutation, N501T, was reported in Brescia (Lombardy) in a single immunocompromised patient (MB61): The same N501T mutation has been observed in mustelids (minks [176, 177] and ferrets [172]): The N501T-G142D variant and N501T-G142D-F486L variant dominate the US mink-derived SARS-CoV-2 sequences [178]. N501Y-specific one-step, real-time RT-PCRs have been developed [179-181].

  Y453F increases affinity to ACE2 (from −12.39 to −10.27 kcal/mol) and partially escapes detection by monoclonal nAbs CC12.1, CC12.3, COVA2-04, CV07-250 [182, 183], etesevimab [131, 184] and casirivimab [184, 185] but not COVA2-39 or CV07-270 [182, 183]. It was the most concerning mutation of the Cluster V variant (discussed below in detail) and also causes evasion from HLA-A24-restricted CTL response [159].

  LF455YL abrogates neutralization by COVA1-12 mAb and reduces that by COVA-2-07 and COVA2-29 mAbs [100].

  TEI470-2NVP prevents neutralization by COVA2-29, COVA2-07 and COVA2-02 mAbs and reduces the activity of COVA1-18 and COVA1-21 mAbs by >100-folds. The neutralization by convalescent sera is only reduced by 2-folds [100].

  A475V is resistant to class I antibodies.

  S477N attenuates neutralization by mAb and convalescent sera [186]. S477N is the hallmark mutation of 20A.EU2 strain [89] (including local variants such as Marseille-4 strain first detected in southern France and Algeria [187].

T478 mutations: **T478K** has been reported in B.1.1.519 from Mexico [188] and in the 20B/B.1.1.222 clade from Mexico, Southern USA and New York [111]. It appeared in 4,214 cases out of 820,000 as of March 26, 2021 [189]. **T478R** characterizes A.VOI.V2 and VOC B.1.617.2.

E484 mutations: **E484K** (nicknamed **"Eeek"**), caused by SNP G23012A, emerged worldwide in March 2020. It is found in more than 64 out of 800 lineages as of March 2021 [190], being a dramatic example of convergent evolution. E.g., it is found in the B.1.351 lineage from South Africa together with N501Y and K417N, in several B.1.1.7 subclades, B.1.1.33(E484K), P.1, P.2, B.1.525, B.1.526, B.1.220 from New York [191], B.1.243.1 from Arizona [192], R.1 from Japan and USA [193] and B.1.1.318 and B.1.621 lineages from UK. The effect of E484K mutation on ACE2 affinity is uncertain, with different reports suggesting either reduced [130], higher [194] or unchanged binding affinity to ACE2 [132]. E484 frequently engages in interactions with antibodies but not with ACE2 [161]. E484K evades neutralization by mAbs and convalescent sera [186] more than 10 folds [123]. In particular, E484K causes resistance to many Class 2 RBD-directed antibodies [195], including DH1041 [196] and bamlanivimab [133]. A majority of the most potent mRNA vaccine-elicited mAbs were at least 10-fold less effective against pseudotyped viruses carrying the E484K mutation [139]. In another study, serum neutralization efficiency was lower against the isogenic E484K rSARS-CoV-2 (vaccination samples: 3.4 fold; convalescent low IgG: 2.4 fold, moderate IgG: 4.2 fold and high IgG: 2.6 fold) compared to USA-WA1/2020 [197]. Another mutation at the same site (**E484Q**) has also been found in a smaller number of human isolates [137, 198] and in VOCs B.1.617.1 and B.1.617.3 [101].

G485R causes ~3 to 5-fold decreases in nAb titer for a few sera [137]. G485 residue does not have a direct interaction with ACE2, its mutation to arginine affects the structure of the 480-488 loop of the RBM, disrupting the interactions of neighboring residues with ACE2 [199].

F490S causes escape to several mAbs [162] and was reported in 20 B.1.1.7 sequences from UK from December 13, 2020 to February 5, 2021 [90].

Q493R causes resistance to class 3 antibodies and especially to both bamlanivimab and etesevimab [200–202].

The occurrence of all the abovementioned mutations in each variant (see section below) is summarized in Table 5.1 and Fig. 5.1. PyR0, a hierarchical Bayesian multinomial logistic regression model that infers relative transmissibility of all viral lineages across geographic regions, detects lineages increasing in prevalence, and identifies mutations relevant to transmissibility.

Table 5.1 Comparison of lineages with regard to mutations in spike and other SARS-CoV-2 genes

Therapeutic/variant		Variants of concern (VOC)			
	PANGOLIN name	B.1.1.7 (Q)	B.1.351	P.1	B.1.617.2 (AY)
	NextStrain name	20I/S:501Y.V1 20I (V1)	20H/S:501Y.V2	20J/S:501Y.V3	21A/S:478K
	PHE name	VOC-20DEC-01	VOC-20DEC-02	VOC-21JAN-02	VUI-21APR02
	WHO name	alpha	beta	gamma	delta
	GISAID name	GRY (formerly GR/501Y.V1)	GH/501Y.V2	GR/501Y.V3	G/452R.V3
	local name	VOC 202012/01, UK variant	501Y.V2 VOC 202012/02	B.1.1.28.1 B.1.1.248 VOC 202101/02	Indian variant
Country of first detection		South-East England, UK	South Africa	Amazonas, Brazil	India
Mutation	ACE2 binding affinity				
V3G	=	–	–	–	–
L5F	=	–	–	–	–
S13I	=	–	–	–	–
L18F	=	– → +	- in V.3 + in v.1. and V.2	– → +	–
T19R	=	.	–	–	+
T20N	=	.	–	+	–
P26S	=	.	–	+	–
H66D	=	.	–	–	–
ΔH69/ΔV70	=	– → +	–	–	–

(continued)

Table 5.1 (continued)

Therapeutic\variant		Variants of concern (VOC)			
V70F	=	–	–	–	+ in AY.2
S71F	=	–	–	–	+
G75V	=	–	–	–	–
T76I	=	–	–	–	–
D80 mutations	=	–	A	A	–
T95I	=	–	–	–	+ in AY.1
D111D	=	–	–	–	+
D138Y	=	–	+	+	–
G142 mutations	=	+	–	–	D
ΔY144	=	+	–	–	–
W152C	=	–	–	–	–
E154K	=	–	–	–	–
E156G	=	–	–	–	+ (– in AY.1)
F157S	=	–	–	–	–
ΔEF156-157	=	–	–	–	+
R158G	=	–	–	+	+
R190S	=	–	+	–	–
R214L	=	–	–	–	–/+ in AY.1
D215G	=	–	+	+	–

(continued)

Table 5.1 (continued)

Therapeutic\variant		Variants of concern (VOC)			
Q218H	=	–	–	–	–
A222V	=	–	–	–	+ in AY.2
Δ242–244	→	–	– → +	– → +	–
R246 mutations	→	–	I in v.1 – in v.2. and v.3	R → I	–
Δ246–252	=	–	–	–	–
ΔSYL247-249	=	–	–	–	–
L249S	=	–	–	–	–
T250I	=	–	–	–	+
D253G	=	–	–	–	–
W258L	=	–	–	–	+ in AY.1
R346K	=	–	–	–	–
K356N	=	–	–	–	–/+
V382L	=	–	–	–	–
P384L	=	–	+ (South Africa)	–	–
K417 mutations	→	–	N	T	N (in AY.1 and AY.2)
N439K	←	–	–	–	–
N440K	←	–	–	–	–
L452 mutations	←	–	–	–	R

(continued)

Table 5.1 (continued)

Therapeutic\variant		Variants of concern (VOC)				
Y453F	↓	–	–	–	–	–
S477N	↓	–	–	–	–	–
T478 mutations	=	–	·	·	·	**K**
V483A	=	–	·	·	·	·
E484 mutations	↓/=	–→**K**	**K**	**K**	**K**	**K**
F490S	↓	–↑+	·	·	·	·
S494P	=	–↑+	+	+	·	·
N501Y	↓	+	+	+	+	+
E516Q	=		+(unclear)			
K558N	=	·	·	·	·	–/+
A570D	=	+				
T572I	=	–		+		+
D614G	↓	+	+	+	+	+
H655Y	↓	+		+	+	+
G669S	=	–				
Q677H	=	–↑+				
P681 mutations	=	H		- →H (Italy)	- →H (Italy)	R
I692V	=					
A701V	=	–	+	+	+	
T716I	=	+	+			
G769V	=	–				+/–

(continued)

Table 5.1 (continued)

Therapeutic variant		Variants of concern (VOC)			
T791I	=	-	-	-	-
D796H	=	-	-	-	-
K854N	=	+/-	-	-	-
T859N	=	-	-	-	-
F888L	=	-	-	-	-
Q949R	=	-	-	-	-
D950 mutations	=	N	-	-	-
S982A	=	-	-	+	+
T1027I	→	-	+	-	-
E1029K	=	-	-	-	-
Q1071H	=	-	-	-	-
Δ1072	=	-	-	-	-
H1101 mutations	=	-	-	-	+
D1118H	=	-	+	+	+
I1130V	=	-	-	-	-
D1139H	=	-	-	-	-
D1153Y	=	-	-	-	-
D1163Y	=	-	-	-	-
G1167V	=	-	-	-	-
V1176F	=	-	− in v.1 + in v.2	-	-
N1187D	=	-	-	-	-

(continued)

Table 5.1 (continued)

Therapeutic\variant	Variants of concern (VOC)				
V1228L	=	-	-	-	-/+
M1229I	=	-	-	-	-
C1248F	=	-	-	-	-/+
C1261F	=	-	-	-	-/+
V1264L	=	-	-	-	-
5'UTR	C241T	+	+	-	-
ORF1ab Δ11288-11296 (Nsp6: Δ3675-3677 SGF)	+	+	+	-	-
ORF1ab other	T1001I, A1708D, I2230T	K1655N	K1655N	K1655N	F106, A488S, P1228L, P1469S, D144, V167L, N244, T492I, T77A, V120, Y87, P323L, G671S, P77L, A394V, R216C
ORF3a	-	-	-	-	S26L
E	-	P71L	P71L	Rarely P71L	-

(continued)

Table 5.1 (continued)

Therapeutic\variant	Variants of concern (VOC)				
M	Rarely I82T, V70L [501]	I82T, V70L [501]	I82T, V70L[501]	I82T, V70L[501]	I82T
ORF6	del2/3 in some E484K sublineages	-	-	-	del2/3 in some AY.2
ORF7a	-	-	-	-	V82A T120I
ORF7b	-	-	-	-	T40I in AY.1 and B.1.617.2
ORF8	Q27stop , R52I, Y73C	-	-	-	ΔDF119
ORF9	-	-	-	-	-
N	D3L, R203K, G204R, S235F AGTAGGG28877-83TCTA AAC (RG203KR)	T205I	P80R T205I	-	D63G R203M G215C (not in AY.2) D377Y
ORF10	-	-	-	-	-
ORF14	-	-	-	-	-
3'UTR	-	-	-	-	-

Therapeutic\variant	Variants of interest (VOI)						
PANGOLIN name	P2	B.1.427/B.1.429	P3	B.1.525	B.1.526	B.1.617.1	C.37
NextStrain name	20B/S.484K	20C/S.452R 21C	20B/S:265C 21E	20A/S484K 21D	20C/S.484K 21F	21A/S:154K 21B	20D 21G
PHE name	VUI-21JAN-01	-	VUI–21MAR-02	VUI-21FEB-03	-	VUI-21-APR-01	VUI-21-JUN-01

(continued)

Table 5.1 (continued)

Therapeutic\variant		Variants of interest (VOI)						
	WHO name	zeta	epsilon	theta	eta	iota	kappa	lambda
	GISAID name	GR	GH/452R.V1	GR	G/484K.V3	GH	G/452R.V3	GR/452Q.V1
	local name	B.1.1.28.2 B.1.1.28(E484K) VUI202101/01	CAL.20C/L452R, West Coast variant	-	Nigerian variant VUI-202102/03 UK1188	-	Indian variant	-
Country of first detection		Rio de Janeiro, Brazil	Southern California, USA	Philippines	Nigeria	New York, USA	India	Peru
Mutation	ACE2 binding affinity							
V3G	=	-	-	-		-	-/+	-
L5F	=	-	-	-	-	+/-	-	-
S13I	=	-	+ B.1.429 − B.1.427	-	-	-	-	-
L18F	=	-	-	-	-	-	-	-
T19R	=	-	-	-	-	-	-	-
T20N	=	-	-	-	-	-	-	-
P26S	=	-	-	-	-	-	-	-
H66D	=	-	-	-	-	-	-	-
ΔH69/ΔV70	=	-	-	-	+	-	-/+	-
V70F	=	-	-	-	-	-	-	-
S71F	=	-	-	-	-	-	-	-
G75V	=	-	-	-	-	-	-	+
T76I	=	-	-	-	-	-	-	+

(continued)

Table 5.1 (continued)

Therapeutic/variant		Variants of interest (VOI)						
D80 mutations	=	–	–	–	–	–	G (B.1.526.1)	–
T95I	=	–	–	–	–	–	+	+ v.1 – v.2
D111D	=	–	–	–	–	–	–	+
D138Y	=	–	–	–	–	–	–	–
G142 mutations		–	–	–	–	–	–	D/del
ΔY144	=	–	–	+	+	+	+ B.1.526.1	–/+
W152C	=	–	+ B.1.429 – B.1.427	–	–	–	–	–
E154K	=	–	–	–	–	–	–	+
E156G	=	–	–	–	–	–	–	–
F157S	=	–	–	–	–	–	+ B.1.526.1	+
ΔEF156-157	=	–	–	–	–	–	–	–
R158G	=	–	–	–	–	–	–	–
R190S	=	–	–	–	–	–	–	–
R214L	=	–	–	–	–	–	–	–
D215G	=	–	–	–	–	–	–	–
Q218H	=	–	–	–	–	–	–	–/+
A222V	=	–	–	–	–	–	–	–
Δ242-244	→	–	–	–	–	–	–	–

(continued)

Table 5.1 (continued)

Therapeutic/Variant		Variants of interest (VOI)					
R246 mutations	→	–	–	–	–	–	–
Δ246–252	=	–	–	–	–	–	+
ΔSYL247-249	=	–	–	–	–	–	–
L249S	=	–	–	¦	–	–	–
T250I	=	–	–	–	–	–	–
D253G	=	–	–	+	–	–	–
W258L	=	–	–	+	–	–	–
R346K	=	–	–	–	–	–	–
K356N	=	.	–	–	–	–	–
V382L	=	.	–	–	–	–/+	–
P384L	=	.	–	–	–	–	–
K417 mutations	→	.	–	–	–	–	–
N439K	←	.	–	–	–	–	–
N440K	←	.	–	–	–	–	–
L452 mutations	←	. → R	–	+ (B.1.526.1)	–	R	Q
Y453F	←	.	–	–	–	–	–
S477N	←	.	.	+ (B.1.526.2)	–	–	–
T478 mutations	=	.	.	.	.	.	.
V483A	=	.	.	.	.	.	.
E484 mutations	↑/=	K	K	-/K	K	Q	.
F490S	←	.	.	.	.	.	+

(continued)

Table 5.1 (continued)

Therapeutic\variant		Variants of interest (VOI)						
S494P	=	·	·	·	·	·	·	·
N501Y	←	·	·	·	+	·	·	·
E516Q	=	·	·	·	·	·	·	·
K558N	=	·	·	·	·	·	·	·
A570D	=	·	·	·	·	·	·	–
T572I	=	·	·	·	·	·	·	–
D614G	←	+	+	+	+	+	+	+
H655Y	←	·	·	·	·	·	·	·
G669S	=	·	·	·	·	·	·	·
Q677H	=	– → + ("Peacock lineage [500]")	·	·	·	·	·	–
P681 mutations	=	H	·	R	·	R	·	–
I692V	=	·	·	·	·	–	–	–
A701V	=	·	·	·	+	–	–	–
T716I	=	·	·	·	·	–	–	–
G769V	=	·	·	·	–/+	·	–	–
T791I	=	·	·	·	–/+	·	·	–
D796H	=	·	·	·	·	·	·	–
K854N	=	·	·	·	·	·	·	–
T859N	=	·	+	·	+/–	+/–	+	+
F888L	=	·	·	·	·	·	·	–
Q949R	=	·	·	·	·	·	·	–

(continued)

Table 5.1 (continued)

D950 mutations	Therapeutic variant	Variants of interest (VOI)				H B.1.526.1	
S982A	=	–	–	–	–	–	–
T1027I	→	–	–	–	–	–	–
E1029K	=	–	+	–	–	–	–
Q1071H	=	–	–	–	–	+	–
Δ1072	=	–	–	–	–	–/+	–
H1101 mutations	=	–	Y	–	–	- in v.1 D/Y in v.2	–
D1118H	=	–	–	–	–	–	–
I1130V	=	–	–	–	–	–	–
D1139H	=	–	–	–	–	–	–
D1153Y	=	–	–	–	–	–/+	–
D1163Y	=	–	–	–	–	–	–
G1167V	=	–	+	–	–	–	–
V1176F	=	–	–	–	–	–	–
N1187D	=	–	–	–	–	–	–
V1228L	=	–	–	–	–	–	–
M1229I	=	–	–	–	–	–	–
C1248F	=	–	–	–	–	–	–
C1261F	=	–	–	–	–	–	–
V1264L	=	–	–	–	–	–/+	–

(continued)

Table 5.1 (continued)

Therapeutic\variant	Variants of interest (VOI)					
5'UTR	C241T	–	–	–	–	–
ORF1ab Δ11288-11296 (Nsp6: Δ3675-3677 SGF)	–	–	D1554G, L3201P, D3681E, L3930F, P4715L, A5692V	–	–	+
ORF1ab other	T10667G (Nsp5: L3468V) C11824T (Nsp6) A12964G (Nsp9)	T265I, I4205V, P314L, D1183Y (+ F2827L, V3367I in Peacock lineage [500])	–	–	–	–
ORF3a	–	Q57H (+A23V in Peacock lineage [500])	–	–	–	–
E	–	–	–	L21F	–	–
M	–	–	–	I82T	–	–
ORF6	–	–	–	del2/3	–	–
ORF7a	–	–	–	–	–	–
ORF7b	–	–	–	–	–	–
ORF8	–	V100L in Peacock lineage [500]	K2Q, R203K, and G204R	-		
ORF9	–	–	–	–	–	–
N	G28628T (A119S) R203K G204R G28975T (M234I)	T205I (+ P142S and M234I in Peacock lineage [500])	R203K G204R	del3 A12G	–	P13L R203K G204R G214C

(continued)

Table 5.1 (continued)

Therapeutic\variant	Variants of interest (VOI)				
ORF10	–	–	–	–	–
ORF14	–	–	–	–	–
3'UTR	–	–	–	–	–

Therapeutic\variant	Variants under monitoring / Alerts for further monitoring				
PANGOLIN name	B.1.258Δ	B.1.1.298	B.1.1.318	B.1.616	AV.1
NextStrain name	–	–	–	–	–
PHE name	–	–	VUI-21FEB-04	–	VUI-21MAY-1
WHO name	–	–	–	–	–
GISAID name	–	–	–	–	–
local name	–	Cluster V	VUI-202102/04	–	–
Country of first detection	Czech Republic, Slovakia	Denmark	England	France	Unclear

Mutation	ACE2 binding affinity					
V3G	=	–	–	–	–	–
L5F	=	–	.	–	–	–
S13I	=	–	–	–	–	–
L18F	=	–	.	.	.	.
T19R	=	–	.	.	.	.
T20N	=	–	.	.	.	.
P26S	=	–	.	.	.	.

(continued)

Table 5.1 (continued)

Therapeutic\variant		Variants under monitoring / Alerts for further monitoring				
H66D	=	–	·	·	+	·
ΔH69/ΔV70	=	+	+	·	·	·
V70F	=	–	–	–	–	–
S71F	=	–	–	–	–	–
G75V	=	–	–	–	–	–
T76I	=	–	–	–	–	–
D80 mutations	=	–	–	–	–	G
T95I	=	–	–	+	–	+
D111D	=	–	–	–	–	–
D138Y	=	–	–	–	–	–
G142 mutations	=	–	–	–	V	D
ΔY144	=	–	–	–	+	+
W152C	=	–	–	–	–	–
E154K	=	–	–	–	–	–
E156G	=	–	–	–	–	–
F157S	=	–	–	–	–	–
ΔEF156-157	=	–	–	–	–	–
R158G	=	–	–	–	–	–
R190S	=	–	–	–	–	–
R214L	=	–	–	–	–	–
D215G	=	–	–	–	+	–

(continued)

Table 5.1 (continued)

Therapeutic/variant	Therapeutic	Variants under monitoring Alerts for further monitoring			
Q218H	=	–	–	–	–
A222V	=	–	–	–	–
Δ242–244	→	–	–	–	–
R246 mutations	→	–	–	–	–
Δ246–252	=	–	–	–	–
ΔSYL247–249	=	–	–	–	–
L249S	=	–	–	–	–
T250I	=	–	–	–	–
D253G	=	–	–	–	–
W258L	=	–	–	–	–
R346K	=	–	–	–	–
K356N	=	–	–	–	–
V382L	=	>	–	–	–
P384L	=	–	–	–	–
K417 mutations	→	–	–	–	–
N439K	←	+	+	–	–
N440K	←	–	–	–	–
L452 mutations	←	–	–	–	–
Y453F	←	+	+	–	–
S477N	←	.	–	–	–
T478 mutations	=	.	–	–	–

(continued)

Table 5.1 (continued)

Therapeutic\variant		Variants under monitoring			Alerts for further monitoring	
V483A	=	·	–	–	+	–
E484 mutations	↑/=	·	–	K	–	K
F490S	←	·	–	>	–	–
S494P	=	·	–	–	–	–
N501Y	←	·	·	·	·	·
E516Q	=	·	·	·	·	·
K558N	=	–	·	·	·	·
A570D	=	·	–	·	–	–
T572I	=	–	–	–	–	–
D614G	←	+	+	+	+	+
H655Y	←	+	–	–	+	–
G669S	=	–	–	+	–	–
Q677H	=	–	–	·	–	–
P681 mutations	=	+	–	·	H	H
I692V	=	–	–	·	–	–
A701V	=	–	–	·	–	–
T716I	=	–	–	·	–	–
G769V	=	–	–	·	–	–
T791I	=	–	–	·	–	–
D796H	=	–	+	·	–	–
K854N	=	–	–	·	–	–

(continued)

Table 5.1 (continued)

Therapeutic\variant				Variants under monitoring Alerts for further monitoring			
T859N	=	−	−	−	−	−	−
F888L	=	−	−	−	−	−	−
Q949R	=	−	+	−	−	−	−
D950 mutations	=	−	−	−	−	−	−
S982A	=	−	−	−	−	−	−
T1027I	→	−	−	−	−	−	−
E1029K	=	−	−	−	−	−	−
Q1071H	=	−	−	−	−	−	−
Δ1072	=	−	−	−	−	−	−
H1101 mutations	=	−	−	−	−	−	−
D1118H	=	−	−	−	−	−	−
I1130V	=	+	−	−	−	−	−
D1139H	=	+	−	−	−	−	−
D1153Y	=	−	−	−	−	−	−
D1163Y	=	−	−	−	−	−	−
G1167V	=	−	−	−	−	−	−
V1176F	=	−	−	−	−	−	−
N1187D	=	−	+	−	−	−	−
V1228L	=	−	−	−	−	−	−
M1229I	=	−	−	−	+	−	−
C1248F	=	−	−	−	−	−	−

(continued)

Table 5.1 (continued)

Therapeutic/variant		Variants under monitoring Alerts for further monitoring		
C1261F	=	–	–	–
V1264L	=	–	–	–
5'UTR		–	–	–
ORF1ab Δ11288-11296 (Nsp6: Δ3675-3677 SGF)		–	–	–
ORF1ab other	Nsp9: M101I, Nsp12: V720I, Nsp13:A598S	T265I, N1324S, T1638I, S2261Y, Y3160H, L3606F, P314L, Q813R, L1681F, T2537I, K2674R	–	–
ORF3a		Q57H	–	–
E		F20L, T30I	–	–
M		H125Y	–	–
ORF6		del23-32 + frameshift resulting in 5 additional aa (HKPHN)	–	–
ORF7a		*122R	–	–
ORF7b		–	–	–
ORF8		–	–	–
ORF9		–	–	–
N		T325I	–	–
ORF10		–	–	–
ORF14		–	–	–

(continued)

Table 5.1 (continued)

Therapeutic:variant	Variants under monitoring			
	Alerts for further monitoring			
	–	–	–	–
3'UTR	–	–	–	–

ORF1ab mutations are represented by its amino acid positions relative to ORF1a (Nsp1-Nsp11) and ORF1b (Nsp12-Nsp16). Modified with permission from reference [481]

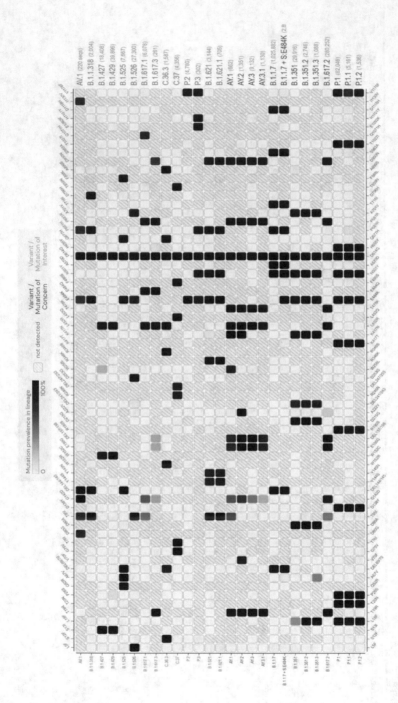

Fig. 5.1. Representation of Spike mutation prevalence in the main VOCs and VOI as of July 14, 2021. Reproduced from Outbreak.info

Chapter 6
SARS-CoV-2 Variants

Abstract This chapter discusses each VOC, VOI, and variant under monitoring.

6.1 B.1.1.298

Minks got infected from humans and back-infected humans [203] in the Netherlands [204, 205], Denmark [183, 206], Canada, Italy, Spain, Sweden, Poland [207,
208] and the USA. In mid-2020, mink-derived variants accounted for 40% of the
total SARS-CoV-2 cases in the Netherlands and were less lethal and transmissive
compared to the native human strains [209]. B.1.1.298 represented one of the Danish
clusters (**Cluster 5 / ΔFVI-Spike**), harboring 4 additional genetic changes (D614G,
I692V and M1229I substitutions, and ΔH69/ΔV70). Despite binding the human
ACE2 receptor with a fourfold higher affinity suggesting an enhanced transmission
capacity, sensitivity to convalescent or vaccine-elicited sera remained unchanged
[210]. Following the lockdown and mass-testing, Danish State Serum Institute (SSI)
announced on November 19, 2020, that cluster 5 in all probability had become
extinct. Cluster V retained robust binding to RBD-directed nAbs, DH1041, DH1043
and DH1047, and improved binding affinity of to the neutralizing NTD-directed
antibodies DH1050.1 and DH1050.2 by 3.5 and 2.6-fold, respectively [196]. On the
contrary, the Utah mink SARS-CoV-2 strain fell into Clade GH, which is unique
among mink and other animal strains sequenced to date and did not share other
Spike RBD mutations Y453F and F486L found in B.1.298 [211]. Another spillover
of mink-adapted SARS-CoV-2 from farmed mink to humans occurred in Poland after
extensive adaptation and lasted at least 3 months, leading to 4 mutations in the S gene
(G75V, M177T, Y453F and C1247F) [208].

6.2   B.1.1.7 (Alpha)

Named **20I/501Y.V1** by NextStrain, **Variant Under Investigation (VUI) or Variant Of Concern (VOC) 202,012/01** by PHE, and **alpha** by WHO, the strain was colloquially known as **UK variant**. It was first reported on December 2020 in England and harbored 14 substitutions and 2 deletions in Spike. The earlier 501Y lineage (occasionally named **501Y variant 1,** which circulated mostly from early September to mid-November) had no ΔH69/ΔV70 and was 10% more transmissible than the 501 N lineage: the later dominant 501Y lineage (ambiguously named **501Y variant 2**, which started circulating at late September) harbored ΔH69/ΔV70, was up to 75% more transmissible than the 501 N lineage [212, 213] and continued to grow during a lockdown in which other lineages shrank [214]. Within households in Norway, an increase in the secondary attack rate by 60% compared to other variants was found. The estimated reproduction number was significantly increased by 24% (0.19 in absolute terms) compared to other variants [215–217]. Compared to the wild-type, B.1.1.7 had comparable surface stability (on steel, silver, copper and face masks) and similar inactivation susceptibility (to heat, soap and ethanol) [218].

B.1.1.7 carries 23 mutations in Spike, ORF8 [219] and N [148]: 7 Spike mutations occur in S1 (ΔH69/ΔV70 and 2 changes in the RBD: N501Y, A570D) and 4 in S2 (P681H, T716I, S982A and D1118H) [48]. Only, the N501Y substitution exhibited consistent fitness gains for replication in the upper airway in the hamster model as well as primary human airway epithelial cells [166]. The non-coding deletion g.a28271-, at upstream of the nucleocapsid (N) gene, triggered the high transmissibility of B.1.1.7. The deletion changes the core Kozak site of the N gene and may reduce the expression of N protein and increase that of ORF9b. The expression of ORF9b is also regulated by another mutation (g.gat28280cta) that mutates the core Kozak sites of the ORF9b gene. If both mutations back-mutate, the B.1.1.7 variant loses its high transmissibility. Moreover, the deletion may interact with ORF1a:p.SGF3675-, S:p.P681H, and S:p.T716I to increase the viral transmissibility [220]. The alpha mutations cause a significant shift in the processing state of N-glycans on one specific N-terminal domain site [221]. B.1.1.7 isolates have dramatically increased subgenomic RNA and protein levels of Orf9b and Orf6, both known innate immune antagonists. Expression of Orf9b alone suppressed the innate immune response through interaction with TOM70, a mitochondrial protein required for RNA sensing adaptor MAVS activation, and Orf9b binding and activity was regulated via phosphorylation [222].

PANGOLIN distinguishes several B.1.1.7 sublineages: • Q.1 • Q.2 • Q.3 • Q.4 harbors P681R (66%) or P681H (32%) • Q.5 harbors D138H (90%) and F490S (84%), and is the rarest substrain • Q.6 • Q.7 • Q.8, on the contrary of Q.1-7, rarely harbors P681H, T716I, and S982A.

- **B.1.1.7 with E484K** (aka **VOC202102-02** or **VOC-21FEB-02** by PHE) was detected in 11 sequences out of 214,159 from Dec 17, 2020, to Jan 26, 2021 [223], which restores the overall ACE2-binding affinity at the same level as N501Y [132]: as of Feb 5, the sequences increased to 27 [90]. The strain has 3.8-folds reduced neutralization by mRNA-1273 and BNT162b2-elicited sera compared to wild-type [224], although another study employing a pseudotyped virus showed no reduction in neutralization by BNT162b2-elicited sera [225].

- Two more RBM mutations leading to potential immune escape were reported: F490S and S494P [90]. Additionally, L18F substitution initiated a substrain characterized by replicative advantage of 1.70 in relation to the remaining VOC-202012/01 substrains [90]. Normalized subgenomic expression profiles are significantly increased in B.1.1.7 infections as a direct consequence of a triple nucleotide mutation in nucleocapsid (28280:GAT > CAT, D3L) creating a transcription regulatory-like sequence complementary to a region 3' of the genomic leader [226].

In addition to SGTF described above, N gene dropout or Ct value shift is specific for B.1.1.7 positive samples using the Allplex SARS-CoV-2/FluA/FluB/RSV PCR assay [227]. B.1.1.7-specific primer sets have been recently designed [228–230], and RT-ddPCR can confirm 4 mutations within the S gene associated with the B.1.1.7 lineage [231].

The strain has similar MR as other lineages: however, B.1.1.7 suddenly appeared with much divergence from the other strains: this can be explained with introduction from a country with poor genomic surveillance, or by regular viral evolution in an animal host before returning to human, or by accelerated viral evolution occurring in a single immunocompromised patient with chronic infection (see paragraph below). Extensive genomic analyses denied in-human evolution and suggested that Canidae, Mustelidae or Felidae could be a possible host of the direct progenitor of variant B.1.1.7 [232]. Alternatively, recombination has been proposed as a mechanism for generation of B.1.1.7 [233]. Accordingly, further recombination has been detected among B.1.1.7 and other strains (B.1.36.17, B.1.36.28, B.1.177, B.1.177.9, B.1.177.16, B.1.221.1): interestingly, in 6 of 8 instances (and all 4 of the transmitting groups), the mosaic viruses contain a Spike gene from the B.1.1.7 lineage and in 4 instances there is a proposed recombination breakpoint at or near the 5' end of the Spike gene [234].

Different reports suggested ACE2-binding affinity by B.1.1.7 as stable [168], 2-folds higher [235], 5.4-folds higher [155] or 10-folds higher [236] when compared to wild-type D614G: the mutations that likely contribute to this phenotype are ΔH69-V70 in the NTD and N501Y in the RBD that enhanced binding by ~1.51 and ~2.52 folds, respectively [155]. Fusogenicity is increased [237]. The replicative advantage of B.1.1.7 has been estimated at 2.24 [238], but, when cultivated separately, viral replication is unchanged in primary human airway epithelial cells and reduced in Vero cells (potentially due to increased furin-mediated cleavage of its Spike protein as a result of a P681H mutation directly adjacent to the S1/S2 cleavage site) [239, 240]. However, when wild-type and B.1.1.7 viruses were put in competition in a human reconstituted bronchial epithelium, B.1.1.7 outcompeted the ancestral strain [241] likely due to enhanced entry [242]. B.1.1.7, compared to an ancestral SARS-CoV-2 clade B virus, produced higher levels of infectious virus late in infection and had a higher replicative fitness in human airway, alveolar and intestinal organoid models [243].

Syrian hamster models suggest increased fitness in upper respiratory tract [244, 245], higher aerosol transmission [246] and strong elevation of proinflammatory cytokines but no increased pathogenesis compared to wild-type strains [247, 248]: previous infection from wild-type strain confers protection [248]. On the contrary a murine model showed 100-folds higher mortality, more severe lesions in internal organs, distinct tissue-specific cytokine signatures, significant D-dimer depositions in vital organs and less pulmonary hypoxia signaling before death compared to wild-type strain: again, previous infection from wild-type strain confers protection [249].

Available SGTF data in community-based diagnostic PCR testing indicate a shift in the age composition of B.1.1.7 reported cases, with a larger share of under 20 years old among reported B.1.1.7 than non-B.1.1.7 cases [250]. Initially, there were no evidence for changes in reported symptoms or disease duration associated with B.1.1.7 [251], but later, studies reported mortality increased by 35% [252, 253]. Another study estimated the adjusted hazard ratios for critical care admission and mortality to be 1.99 and 1.59 for VOC B.1.1.7 compared with the original variant group, respectively; the adjusted hazard ratio for mortality in critical care was 0.93 for patients with VOC B.1.1.7, compared to those without [254]. The proportion of cases with hypoxia on admission was greater in those infected with the B.1.1.7 variant [255].

HLA-A2$^+$ CD8$^+$ T cell epitopes from B.1.1.7 had lower binding capability than those from the ancestral strain. In addition, these peptides could effectively induce the activation and cytotoxicity of CD8$^+$ T cells. At least, two site mutations in B.1.1.7 resulted in a decrease in CD8$^+$ T cell activation and a possible immune evasion, namely A1708D mutation in ORF1ab1707-1716 and I2230T mutation in ORF1ab2230-2238 [256].

Sera from persons vaccinated with BNT162b2 neutralized isogenic Y501 SARS-CoV-2 strain (generated on the genetic background of the N501 clinical strain USA-WA1/2020) [257] or B.1.1.7 Spike pseudotypes (Δ69–70 + N501Y + A570D [258–260] or the full set of mutations [261]) or authentic B.1.1.7 [239, 262–265] with equivalent or less than threefold reduced titers compared to wild-type strain. No impact was detected on neutralization titers when using sera from human subjects vaccinated with mRNA-1273 [266–268] or COVAXIN [269]. In a large study in UK, no HCW vaccinated twice with BNT162b2 or ChAdOx1 had symptomatic infection, and incidence was 98% lower in seropositive HCWs. Two vaccine doses or seropositivity reduced the incidence of any PCR-positive result with or without symptoms by 90% and 85%, respectively. Single-dose vaccination reduced the incidence of symptomatic infection by 67% and any PCR-positive result by 64% [270].

Overall, B.1.1.7 causes resistance to neutralization by the NTD–specific neutralizing mAbs [271], such as COVA2-17, COVA1-12 and COVA1-21 [267], but not 1–57, 2–7 [272] and bamlanivimab [264]. Antibodies biding the NTD antigenic supersite are ineffective, while antibodies-binding RBD and fusion peptide (DH1058) remain effective[196]. Most convalescent human sera showed neutralization reduced by <3-folds [100, 239, 262, 267, 268] and hamster models confirm protection against B.1.1.7 from previous infection [273]. Accordingly, only a single patient (previously infected with B.2) has been reported as getting genotypically defined B.1.1.7 reinfection to date in UK, despite intensive genomic monitoring [274], and 2 in Italy [275] and 1 in USA [276]. The reinfection rate 0.7% is similar to older strains [251]. Co-infection between B.1.1.7 and B.1.351 has been reported in 90-years-old female from Belgium [277].

The mutations seen in B.1.1.7 would not result in loss of dominant nAb responses to linear Spike glycoprotein and nucleoprotein epitopes in the vast majority of COVID patients [278]. Antibodies elicited by B.1.1.7 infection exhibited significantly reduced to undetectable neutralization of parental strains, B.1.351 [279] or B.1.427/B.1.429 [280] VOCs compared to that of B.1.1.7, which is different from what was observed with B.1.351-elicited sera.

B.1.1.7 first appeared on Sep 20 in South-East England [281] but soon spread detected across all continents and became dominant in Europe within March 2021 [282]. Facebook mobility data (https://visualization.covid19mobility.org/?region=WORLD predicted the spreading of B.1.1.7 [283] across all continents (https://cov-lineages.org/global_report_B.1.1.7.html) [284, 285]. B.1.1.7 spread with greater transmission in colder and more densely populated parts of England and had transmission advantage at warmer temperatures versus other strains, implying that Spring conditions facilitated B.1.1.7's invasion in Europe and across the Northern hemisphere [286].

6.3 B.1.351 (Beta)

Named **VOC 202012/02** or **VOC-20DEC-02** by PHE, **20H/501Y.V2** by NextStrain and **beta** by WHO, this lineage was known colloquially as South African variant. It was found since October 2020 in Nelson Mandela Bay in the Eastern Cape Province of South Africa and spread across all continents (https://cov-lineages.org/global_rep ort_B.1.351.html) [284]. The strain harbored D80A, D215G, ΔL242-A243-L244, K417N, E484K, N501Y, and A701V as a signature [105] and evolved from GISAID clade GH [287]. K417N and E484K reduce the ACE2-binding affinity by abolishing 2 interfacial salt bridges that facilitate RBD binding, K417(S)-D30(ACE2) and E484 (S)-K31(ACE2). These two mutations may thus be more than compensating the attractive effect induced by N501Y, overall resulting in an ACE2-binding affinity variably reported as unchanged [288] or fivefold higher than wild-type [235], i.e., threefold stronger than wild-type RBD but threefold weaker than N501Y [132]. The NTD substitution R246I decreased ACE2-Fc binding by ~1.52 folds, the Δ242-244 deletion by ~1.35 folds, whereas K417N had a greater impact with a decreased binding of ~7.7 folds relative to D614G [155]. Viral fusogenicity is increased [237]. PANGOLIN distinguishes 4 substrains:

- **B.1.351.1** (aka **B.1.351 v.1**), first reported in Botswana, additionally harbors L18F and R246I but does not harbor ΔL242-A243-L244, E484K nor N501Y
- **B.1.351.2** (aka **B.1.351 v.2**), first reported in Mayotte, additionally harbors L18F
- **B.1.351.3** (aka **B.1.351 v.3**), first reported in Bangladesh, harbors no additional mutations

- **B.1.351.4**

B.1.351 presented an intermediate viral load in nasopharyngeal swabs between the B.1.1.7 and wild-type [289], and it is found more easily in saliva. Compared to wild-type, the strain has comparable surface stability on steel, silver, copper and face masks, and similar inactivation susceptibility to heat, soap and ethanol [218]. The strain in cell culture grows toward the higher end of the variants [240].

K417N and E484K abolished the salt bridges between Spike and casirivimab [290], BD23, H11_H4, and C105 [291], but not VH-Fc ab8 [292], 1–57 and 2–7 [272]. Hamster models suggested no increased pathogenesis compared to wild-type strains [247]. The strain was fully resistant to bamlanivimab [264, 290, 293], CA1, etesevimab and CC12.1, and, most importantly, convalescent sera were no longer neutralizing in 48% of cases (only 7% retaining $ID_{50} > 400$) [262, 267, 290, 294–296]. Despite loss of recognition of immunodominant CD4 epitope(s) and 12.7 fold reduction in neutralization of a B.1.351-pseudotyped lentivirus by serum from convalescent subjects, overall $CD4^+$ and $CD8^+$ T cell responses to B.1.351 are preserved [297].

Hamster models suggested protection from B.1.351 after previous infection [273]. A murine model showed 100-folds higher mortality, more severe lesions in internal organs, distinct tissue-specific cytokine signatures, significant D-dimer depositions in vital organs and less pulmonary hypoxia signaling before death compared to wild-type strain: again, previous infection from wild-type strain confers protection [249]. K18-hACE2 transgenic mice challenged with the B.1.351 variant displayed a faster progression of infection. Furthermore, B.1.351 can establish infection in wild-type mice [298], while B.1 cannot. B.1.351-challenged wild-type mice showed a milder infection than transgenic mice [299]. In humans, compared to Alpha (B.1.1.7) variant, odds of progressing to severe disease were 1.24-fold higher, odds of progressing to critical disease were 1.49-fold higher, and odds of COVID-19 death were 1.57-fold higher for Beta [300].

One case of reinfection from B.1.351 (4 months after non-B.1.351) has been documented to date [301] and 4 in Luxembourg [276]. Co-infection with B.1.1.7 and B.1.351 has been reported in a 90-years-old patients from Belgium [277]. Sera from B.1.351-infected patients maintained good cross-reactivity against viruses from the first wave and P.1 VOC [302]. Adaptive mutations in the E gene might have had associated fitness costs that were subsequently recouped by secondary mutations elsewhere in the gene [303]. B.1.351-specific primer sets have been recently designed [229].

mRNA-1273-elicited sera led to 2.7 and 6.4-fold geometric mean titer (GMT) reduction (albeit still 1:190) in neutralization against K417N + E484K + N501Y + D614G or full B.1.351 Spike pseudovirus, respectively, when compared to the D614G VSV pseudovirus [266, 295]. Similarly, BNT162b2-elicited sera led to 0.81-to 1.46-fold GMT reduction in neutralization against a E484K + N501Y + D614G Spike pseudovirus [259, 260, 290] or authentic B.1.351 [166, 262–265], although still 1:500, a titer that was higher than the average titer with which convalescent sera neutralized D614G. Binding of a triple mutant (E484K, K417N, N501Y) RBD to

BN162b2 vaccinee sera was reduced by 10-folds [304]. Immunization with a single dose of mRNA-1273 or BNT162b2 vaccine generated a 1000-fold increase in nAb titers against B.1.351 [305]. And finally, sera from persons vaccinated with one of 2 Chinese vaccines (BBIBP-CorV or recombinant dimeric receptor-binding domain (RBD) vaccine ZF2001) largely preserved neutralizing titers, with slightly reduction, against B.1.351 authentic virus [306]. Unfortunately, a two-dose regimen of ChAdOx1-nCoV19 vaccine did not show protection against mild-moderate COVID-19 due to B.1.351 [307]. Initial infection with SARS-CoV-2 lineage A in hamsters does not prevent heterologous reinfection with B.1.351 but prevents disease and onward transmission [308].

6.4 B.1.1.28- and B.1.1.33-Derived Brazilian Variants (Including Gamma and Zeta)

A task force created by the scientific community was able to sequence only approximately 0.03% of all positive SARS-CoV-2 cases through the pandemic's first year [309, 310]. As of March 2021, 59 different lineages were announced in Brazil, being the majority sequenced in São Paulo, Rio de Janeiro, Rio Grande do Sul, and Amazonas [309]. Several VOIs and VOCs have been reported:

- derived from B.1.1.28: the original B.1.1.28 lineage emerged in Brazil as soon as February 2020 [63, 105, 311]. At least, 5 child variants have been identified [312]:

  - **P.1 VOC** (named **gamma** by WHO, **20 J/501Y.V3** by NextStrain, **VOC-21JAN-02** by PHE or improperly termed **B.1.1.28.1 or B.1.1.248,** or **VOC 202101/02**) was first reported in January 2021 in 4 Japanese travelers returning from Manaus, the capital of Amazonas state in northern Brazil [313]. Such area had a 76% seroprevalence at October 2020 after a largely unmitigated first wave [314], but P.1 was able to cause 4 times more cases during a major second wave beginning in December 2020 [187], which caused significant increases in CFR in young and middle-aged adults [315]. P.1 transmissibility is about 2.5 times higher compared to the previous variant in Manaus, and a low probability of reinfection by P.1 (6.4%) was estimated [316, 317]. One case of reinfection has been documented months after B.1 primoinfection [318]. A serosurvey in blood donors at May 2021 showed that (assuming that reinfections induce a recrudescence (or boosting) of plasma anti-N IgG antibody levels, yielding a V-shaped time series of antibody reactivity levels), 16.9% of all presumed P.1 infections that were observed in 2021 were reinfections [319]. The clade later spread to Rio Grande do Sul [320] and finally led to imported cases [321] and clusters [322] worldwide (https://cov-lineages.org/global_report_P.1.html). P.1 harbors E484K, K417N and N501Y: such combination induces conformational change greater than N501Y mutant alone [323]. The Spike from P.1 presents a ~ 4.24-fold increase in binding compared to D614G: few NTD mutations, namely T20N, P26S, D138Y and R190S, likely contributed to the increase in ACE2 binding, with ~2, ~1.6, ~1.3 and ~1.8-fold increase compared to D614G, respectively. Interestingly, the RBD mutation K417T and the S2 mutation T1027I decreased the ACE2-Fc by ~1.3 and ~1.7 folds respectively. The H655Y mutation, near the S1/S2 217 cleavage site, also slightly increased ACE2 interaction by ~1.2 folds [155]. P.1-specific primer sets have been recently reported [229]. Based on similarity with cluster V RBD, P.1 was predicted to be highly resistant to both etesevimab and casirivimab [184]: retained sensitivity to imdevimab [324], partial resistance to casirivimab [290, 324] and full resistance to bamlanivimab [290, 293, 324] and etesevimab [324] were later confirmed using P.1 pseudovirus, as well as resistance to 4 more NTD antibodies (2–17, 4–18, 4–19, and 5–7) [324], but not the other 2 mAbs 5–24 and 4–8 targeting the antigenic supersite in NTD [324]. P.1 is also more resistant to neutralization by CCP (6.5 fold) and vaccine-elicited sera (2.2–2.8 fold) [324]: binding of a triple mutant (E484K, K417N, N501Y) RBD to BNT162b2 vaccinee sera is reduced by 10-folds [304], while binding of the authentic Victoria strain is reduced by 2.6 folds [325]. Binding of ChAdOx1 vaccine-elicited sera was reduced by 2.9 folds [325]. P.1 may be 1.4–2.2 times more transmissible and 25–61% more likely to evade protective immunity elicited by previous infection with non-P.1 lineages [326]: increase in the risk of severity and fatality rate from P.1 was greater among young adults without preexisting risk conditions [327]. Resistance to convalescent o vaccinee sera was confirmed in murine models [328]. Co-infection by B.1.1.248 (either as major or minor haplotype) and B.1.1.33 or B.1.91, respectively, has been reported [310]. Unlike wild-type, P.1 is able to infect common laboratory mice, replicating to high titers in the lungs [329]. 84% sera from subjects who had been previously asymptomatically infected with B.1.1.28 contained nAbs against the ancestral and P.1 strains, respectively, and remained positive throughout the 6-week study period: neutralization titers were consistently higher against the ancestral strain, but in the majority of patients (57.8%), differences did not differ by more than a single dilution [330]. Convalescent P.1 patients are less protected from other SARS-CoV-2 strains with an important reduction of nAbs against 20A.EU1 and B.1.1.7, about 12.2 and 10.9-fold, respectively [331]. VOC Gamma induces higher viral loads (N1 target; mean reduction of Ct: 2.7) [332]. E661D in Spike protein has been identified in nearly 10% of the genomes from Paraná in March and April 2021 [333]. PANGOLIN distinguishes several subclades:

P.1 v.1 harbors no additional mutations

Gamma plus (P.1+) harboring 2 types of additional Spike changes, having a sharp increase (78% in the second half of May 2021): deletions in the N-terminal (NTD) domain (particularly Δ144 or Δ141-144) associated with resistance to anti-NTD nAbs or mutations at the S1/S2 junction (N679K or P681H) that probably enhance the binding affinity to the furin cleavage site [357]. As lineages P.1.4 (S:N679K) and P.1.6 (S:P681H) expanded (Re > 1) from March to July 2021, the lineage P.1 declined (Re < 1) and the median Ct value of SARS-CoV-2 positive cases in Amazonas significantly decreases. Still, we found no overrepresentation of P.1+ variants among breakthrough cases of fully vaccinated patients (71%) in comparison to unvaccinated individuals (93%).

P.1 v.2 additionally harbors V1176F, and it has been recently described in Porto Alegre and more Brazilian states and foreign countries [334]. Another P1-related sublineage harboring mutation L106F in ORF3a is represented by 8 genomes from the Tocantins [335].

a genomic survey in the Amazonas revealed a sharp increase (78% in the second half of May 2021) in the prevalence of P.1 variants harboring deletions in the NTD domain (particularly Δ141-144) or mutations at the S1/S2 junction (N679K or P681H) [336].

- **P.1-like-II** lineage genomes share some, but not all, defining mutations of the VOC P.1. For instance, it has the previously described ORF1a:D2980H and N:P383 L unique mutations and the newly detected ORF1a:P1213L and ORF1b:K2340N mutations [333].
- **P.2** was also named **VUI-21JAN-01** (formerly **VUI-202101/01**) by PHE, or improperly **B.1.1.28.2** or **B.1.1.28(E484K)**). It was classified as VOI **zeta** by WHO on March 17, 2021, and then reclassified as Alert for further monitoring on July 6, 2021. P.3 was first reported in Rio de Janeiro, harboring E484K as the lone Spike mutation, plus 5 mutations in the UTRs, ORF8 and N: since the first report, two more mutations in orf1ab (U10667G > L3468V and C11824U > I3853I) were reported since December 2020 [105, 311]. At least, 2 cases of reinfection have been documented months after B.1.1.33 primoinfection [337, 338]). P.2 was also detected in the northeast region of Brazil in the states of Bahia and Rio Grande do Norte [338]. P.2 is fully resistant to bamlanivimab but only slight resistant to BNT162b2-elicited sera compared to wild-type strain [264]. P.2 induced body weight loss, viral replication in the respiratory tract, lung lesions and severe lung pathology in infected Syrian hamster model in comparison with B.1: sera from P.2 infected hamsters efficiently neutralized the D614G variant virus, whereas sixfold reduction in the neutralization was seen in case of D614G variant infected hamsters sera with the P.2 variant [339].
- **P.4** (originally known as **VUI-NP13L**), characterized by 12 lineage-defining mutations [310].

P.4.1 presents 4 additional unique amino acid changes in ORF1a (P22875S, V2588F, L3027F, Q3777H) and two synonymous mutations in the ORF1a gene (C1288T and G10870T) [340]. P.4.1 probably emerged in Goiás or São Paulo around Jun-Jul 2020, but this lineage was only identified in South Brazil at the beginning of October 2020. According to an independent phylodynamic reconstruction, P.4.1 rapidly arrived in the southeastern and northeastern regions of Brazil and seems to have been exported to Japan, the Netherlands and England [312]

- 2 sequences (LBI215 and LBI218) characterized by Spike E484Q and N501T, in addition to L18F, Δ142–143 GV, L938F, D950N, C28311U [341]
- **P.X** presents additionally non-synonymous mutations when compared to B.1.1.28, including N234P and E471Q in S protein [342]
- **B.1.1.28 + Q675H + Q677H** clade probably arose around November 2020, in Montevideo, the capital department of Uruguay. This clade further spread from Montevideo to other Uruguayan departments, with evidence of local transmission clusters in Rocha, Salto and Tacuarembo. It also spread to the USA and Spain. The non-synonymous mutations in the viral Spike Q675H and Q677H, which are among the 10 lineage-defining mutations, are in the proximity of the polybasic cleavage site at the S1/S2 boundary and also are reported as recurrent arising independently in many SARS-CoV-2 lineages circulating worldwide by the end of 2020. Although the B.1.1.28 + Q675H + Q677H lineage was later substituted by the VOC P.1 as the most prevalent lineage in Uruguay since April 2021 [343, 344].

- derived from B.1.1.33:

  - E484K has also been found in B.1.1.33 lineage from São Paulo and Amazonas, and has been termed **B.1.1.33(E484K)** [345] or **N.9 variant of interest** [346], harboring the additional mutations NSP3:A2529V (A1711V), NSP6:F3605L(F36L), and NS7b:E33A.
  - A related B.1.1.33-derived lineage, termed **N.10**, harbors 14 lineage-defining mutations. It displays as the most remarkable genetic changes the V445A and E484K mutations in the S protein RBD, several non-synonymous mutations (P9L, I210V, and L212I), and three deletions (Δ141–144, Δ211 and Δ256–258) in the S protein NTD, including a truncated NS7b protein due to a frame-shifting deletion. Other 5 lineage-defining mutations were found among NSP3, NSP5, NSP6 and N. This VOI probably emerged in late December 2020 and comprises a significant fraction (23%) of the SARS-CoV-2 positive cases detected in the Brazilian state of Maranhão (Northeastern region), between January and February 2021 [347]. S:W152C was later reported in one genome from Paraná [333].

6.5   B.1.525 (Eta)

Named **eta** by WHO, **VUI-21FEB-03** (previously **VUI-202102/03**) by PHE, **20A/S484K** by NextStrain, and formerly known as **UK1188**, this lineage harbors E484K, ΔH69/ΔV70, and a F888L in Spike, I82T in M, and del2/3 in ORF6. As of March 5, it had been detected in 23 countries.

6.6   B.1.526 (Iota)

Named **iota** by WHO and **20C/S.484K** by NextStrain, this variant, harboring E484K, was discovered in New York City in November 2020. As of April 11, 2021, the variant has been detected in at least 48 US states and 18 countries. In a pattern mirroring B.1.429 of California, B.1.526 was initially able to reach relatively high levels in some states, but in the Spring of 2021, it was outcompeted by the more transmissible B.1.1.7. Spike from B.1.526 showed a ~1.8-fold increase over D614G [155]. Iota can cause vaccine breakthrough infections [348]. PANGOLIN distinguishes several subclades:

- **B.1.526.1**
- **B.1.526.2** harbors L5F, T95I, D253G, E484K, D614G and A701V and was reported in northeast of the USA [349, 350]
- **B.1.526.3**

6.7   B.1.427/B.1.429 (Epsilon)

Named **20C/S.452R** or **21C** by NextStrain, or **CAL.20C**, this lineage was first named as **epsilon** VOI by WHO on March 5, 2021, and reclassified as Alert for further monitoring on July 6, 2021. It emerged around May 2020 in California and increased from 0 to >50% of sequenced cases from September 1, 2020, to January 29, 2021, exhibiting an 18.6–24% increase in transmissibility relative to wild-type circulating strains. The variants share the L452R substitution (which increases ACE2-Fc binding by ~2.7 221 folds), while S13I and W152C (which do not affect ACE2 affinity) only occur in B.1.429. Spike from B.1.429 augmented ACE2-Fc interaction by ~2.8 folds [155]. It has to twofold increased viral shedding in vivo and increased L452R pseudovirus infection of cell cultures and lung organoids. Antibody neutralization assays showed 4.0 to 6.7-fold and 2.0-fold decreases in neutralizing titers from convalescent patients and vaccine recipients, respectively [104]. While the WHO classifies them as a VOI (epsilon), the CDC classifies them as a VOC.

6.8   B.1.617-Derived Variants (Including Kappa and Delta)

It was first detected on October 5 2020 in Maharashtra, India, and jumper to 80% prevalence on April 1, overwhelming B.1.1.7 [101]. There are 3 sublineages of B.1.617, which have some differences in their exact mutations, but all sharing (in addition to D614G) *L452R and P681R* and, albeit at varying frequencies, *G142D* in Spike and R203M and D377Y in N.

- B.1.617.1 (initially known as B.1.617, but later renamed **kappa** by WHO, **21A/S.154K** in NextStrain, or **VUI-21APR-01 by PHE**) additionally has the

E154K, E484Q and Q1071H mutations. Hamsters infected with B.1.617.1 have more body weight loss, more lung lesions and hemorrhage compared to B.1 wild-type [351]. B.1.617.1 Spike have lower ACE2 affinity than D614G Spike [155].

- **B.617.1 v.1** additionally harbors *T95I*
- **B.617.1 v.2** additionally harbors *H1101D*

- Delta VOC actually consists of 4 sublineages sharing additional *T19R, del157/158, T478K, and D950N* in Spike and I82T in M. The receptor-binding beta-loop-beta motif adopts an altered but stable conformation causing separation in some of the antibody-binding epitopes [352].

  - **B.1.617.2** (aka **VUI-21APR-02** by PHE) is the references strain of Delta. del157/158 in NTD of Spike lead to immune evasion through antibody escape [353]. B.1.617 entered 2 out of 8 cell lines tested (including Calu-3 lung cell [354]) with increased efficiency. K77T and T95I (typical of B.1.617 v.1), L216F, A222V, and G1124V have also been rarely reported. B.1.617.2 is resistant to bamlanivimab [354, 355] and moderately evades convalescent or BNT162b2-elicited sera [354, 355], that is similar in magnitude to the loss of sensitivity conferred by L452R or E484Q alone. Furthermore P681R mutation significantly augments syncytium formation in Calu-3 cells [354] and hamsters [356] compared to the B.1.617.1 Spike protein, potentially contributing to increased pathogenesis observed in hamsters and infection growth rates observed in humans [356]. In hamsters, higher shedding of subgenomic RNA has been reported for 14 day, and moderate lung pathology has been reported in 40% of infected animals [357]. It can cause vaccine breakthrough events [358] and has been reported to infect Asiatic lions (*Panthera leo persica*) in Arignar Anna Zoological Park, Chennai, Tamil Nadu, India [359]. Analysis of an outbreak involving 167 cases from China showed that the time interval between exposure to first positive PCR was shorter (4 vs. 6 days), the initial viral load 1260-folds higher and more infectious (80% vs. 19% harboring $> 6 \times 10^5$ viral copies/ml) than in 19A/19B infection: some minor intra-host single nucleotide variants (iSNVs) could be transmitted between hosts and finally fixed in the virus population during the outbreak. The minor iSNVs transmission between donor-recipient contribute at least 4 of 31 substitutions identified in the outbreak suggesting some iSNVs more likely to arise and reach fixation when the virus spread rapidly [360]. A larger study from Netherlands showed that Delta variant NPS had about fourfold higher viral loads compared to the non-VOC or Alpha variants [361]. Neutralization by CCP, BNT162b2, mRNA-1273 and Ad26.COV2.S-elicited antibodies is reduced by 3–5 folds, while REGN10933 efficacy is reduced by 12 folds compared to wild-type (having only a minor effect on the activity of the REGN-COV2 cocktail) [362]. PANGOLIN distinguishes 33 Delta sublineages termed AY.1 to AY.33. AY.1 and AY.2 harbor K417N and were initially colloquially referred as "Delta plus".

– **AY.1** (aka **B.1.617.2.1**) was first reported in India on April 2021 and spread to 20 countries as of July 2021 (expecially Portugal, Japan, USA, UK and Switzerland) [363]. It additionally harbors *W258L and K417N* (as seen in B.1.351). R214L and K558N and S1261F have also been rarely reported. It also harbors mutations in the genome at ORF1a (A1306S, P2046L, P2287S, V2930L, T3255I, T3646A), ORF1b (P314L, G662S, P1000L, A1918V), ORF3a (S26L), M (I82T), ORF7a (V82A, T120I), ORF7b (T40I), ORF8 (del119/120) and N (D63G, R203M, G215C, D377Y). Neutralization by CCP, BNT162b2, mRNA-1273 and Ad26.COV2.S-elicited antibodies is reduced by 3–5 folds, while REGN10933 efficacy is reduced by 92.7 folds compared to wild-type (having only a minor effect on the activity of the REGN-COV2 cocktail) [362].
– **AY.2** additionally harbors *V70F, E156G, A222V and K417N*. K356N and V1228L have also been rarely reported
– **AY.3** additionally harbors *E156G*

> **AY.3.1** never harbors T95I nor A222V and harbors F157C and R158G at about 10% frequencies

• **B.1.617.3** (aka **VUI-21APR-03** by PHE) has the E484Q mutation and, despite its name, was the first sublineage of this variant to be detected, in October 2020 in India. It was initially very improperly known as "**double mutant**," additionally harboring *T19R, del157/158, E484Q, D950N and E1072K*. V6F, A27V, del69-70, F79S, del142, del144-145, H655Y, Q779K, del950 and H1101D are also rarely reported in Spike. Mutations at L452R and E484Q increased the stability and intra-chain interactions in the Spike protein, which may change the interaction ability of human antibodies [364]. It also harbors P67S in N. This sublineage has remained relatively uncommon compared to the two other sublineages, B.1.617.1 and B.1.617.2, which were both first detected in December 2020. The unprecedented growth of B.1.617.2 cases in India occurred in the background of high seropositivity, but with low median nAb levels, in a serially sampled cohort: vaccination breakthrough cases over this period were noted, disproportionately related to VOC in sequenced cases, but usually mild [365].

Out of total 38 identified mutations among Indian SARS-CoV-2 Nsp13 protein, four mutant residues at position 142 (E142), 245 (H245), 247 (V247) and 419 (P419) are localized in the predicted B cell epitope region [366].

6.9   C.37 (Lambda)

Initially classified as a B.1.1.1 sublineage, it was designated VOI **lambda** by WHO on June 14, 2021. It presents a deletion in the ORF1a gene (Δ3675-3677), also present in B.1.1.7, B.1.351 and P.1. It displays a novel deletion and multiple non-synonymous mutations in the Spike gene (ΔRSYLTPG246-252, G75V, T76I, L452Q, F490S, T859N), and I82T in M. A subvariant (PV29369) has been reported which contained additional changes compared to the consensus sequence including a large 13 amino acid deletion (ΔT63:G75) instead of G75V and T76I NTD substitutions and an additional E471Q substitution in the RBD [224]. T76I and L452Q mutations cause higher infectivity, while the RSYLTPGD246-253N deletion is responsible for evasion from nAbs [367]. Initially reported in Lima, Peru, in late December 2020, it accounted for 48% of genomes in Lima between January 1 and March 18, 2021. Further RT-qPCR screening for VOCs suggests that it is widespread in other regions of Peru. Many imported cases from Peru have been reported [368]. It is also expanding in Chile and Argentina, and there is evidence of onward transmission in Colombia, Mexico, the USA, Germany, and Israel [369, 370]. Neutralization by CCP, BNT162b2 [224, 362], mRNA-1273 [224, 362] and Ad26.COV2.S [362]-elicited antibodies is reduced by 3–5 folds [362], while REGN10933 and REGN10387 efficacy is minimally reduced compared to wild-type [362].

6.10 P.3 (Theta)

Also named **B.1.1.28.3, VUI-21MAR-02** by PHE, **PHL-B.1.1.28**, or **21E** by NextStrain or **GR/1092K.V1** by GISAID, it was designated VOI theta by WHO on Mar 24, 2021 and then reclassified as Alter for further monitoring on July 6, 2021. It was first reported from the Central Visayas region of the Philippines, and imported cases were reported in the USA [371] and elsewhere. This emergent variant is characterized by 13 lineage-defining mutations, including ΔLGV141-143, E484K, N501Y, D614G, P681H, E1029K, H1101Y and V1176F in Spike, D1554G, L3201P, D3681E, L3930F, P4715L, A5692V in ORF1ab, and K2Q, R203K, and G204R in ORF8 [372, 373].

6.11 Other Variants Under Monitoring (VUM)

Many more variants are being reported with increasing sequencing efforts worldwide, some of them having meaningful mutations. They are variably referred as "variants under monitoring (VUM)" by ECDC or "Alerts for further monitoring" by WHO. Examples include:

- **AV.1** (aka **VUI-21MAY-01** in PHE, and belonging to GISAID clade GR) was first reported from South Yorkshire, UK, in March 2021, harboring 23 additional mutations (including D80G, T95I, G142D, Δ144, N439K, E484K, D614G, P681H, I1130V, D1139H) when compared to the majority of other samples of the parent lineage B.1.1.482 [374]. It also harbors A63T and H125Y in M, and I157V in N. It was designated an Alert for further monitoring on May 26, 2021.
- a new variant in West Bengal, India, which is characterized by the presence of 11 co-existing mutations including D614G, P681H and V1230L in S-glycoprotein [375].

- **A.VOI.V2** from Angola harboring 31 amino acid substitutions. The 11 Spike mutations include 3 substitutions in the RBD (R346K, T478R and E484K); 5 substitutions and 3 deletions in the N-terminal domain, some of which are within the antigenic supersite (ΔY144, R246M, ΔSYL247-249 and W258L); and 2 substitutions adjacent to the S1/S2 cleavage site (H655Y and P681H) [376]. It reduces neutralizing activity of mRNA-1273-elicited nAb by 8-folds [377].
- **A.23** viral lineage (belonging to NextStrain clade 19B), is characterized by three Spike mutations F157L, V367F and Q613H, was first identified in a Ugandan prison in July 2020, and then spilled into the general population adding additional Spike mutations (L141F and P681R) to comprise lineage **A.23.1** by September 2020—with this virus being designated a VOI in Africa and with subsequent spread to 26 other countries [115].

  - **A.23.v1** preserves neutralization by mRNA-1273-elicited sera [224, 377]
  - **A.23.v2** also includes R102I and E484K

- **A.27.RN** from Germany comprising L18F, L452R, N501Y, A653V, H655Y, D796Y and G1219V with a later gain of A222V. It emerged in parallel with the B.1.1.7 variant, increased to >50% of all SARS-CoV-2 variants by week 5 of 2021. Subsequently, it decreased to <10% of all variants by calendar week 8 when B.1.1.7 had become the dominant strain. Antibodies induced by BNT162b2 vaccination neutralized it with a 2-threefold reduced efficacy as compared to the wild-type and B.1.1.7 strains [378]
- **AP.1** (https://cov-lineages.org/lineages/lineage_AP.1.html), aka the "**Wales lineage**" descending from **B.1.1.70** within clade **20B**. With already 23 mutations difference to the Wuhan reference virus, the lineage additionally evolved the Spike mutation N501Y and was subsequently predominantly detected in Wales. The mutation is estimated to have emerged around Aug 2020. Mobility of the virus to Saxony, Germany, can be detected, coinciding with an additional mutation in ORF3a:V259A [91].
- **AT.1** (belonging to GISAID clade GR) was reported from Russian Federation on January 2021 and was designated as Alert for further monitoring by WHO on June 9, 2021
- **B.1.x** (including **B.1.342.1**) from California harboring Spike mutations S494P, N501Y, D614G, P681H, K854N, and E1111K and N:M234I (G28975A), which also appears in Variants of Interest B.1.526 and P.2 (G28975T). Of interest, a 35-bp deletion in ORF8 causes the sequence to be automatically rejected from both GISAID and GenBank repositories [379].
- **B.1.111** from Colombia harboring L249S and E484K [190]
- **B.1.177.637.V2/20E (VOI1163.7.V2)** from Spain harboring E484K and Δ141–144 in addition to D1163Y and G1167V which characterize VOI1163.7.V1 [380].
- **B.1.214.2** (belonging to GISAID clade G) first appeared in multiple countries in November 2020 and was designated as Alert for further monitoring by WHO on June 30, 2021
- **B.1.362 + L452R** variant demonstrated a X4-fold reduction in neutralization capacity of sera from BNT162b2-vaccinated individuals compared to a wild-type

strain. The variant infected 270 individuals in Israel between December 2020 and March 2021, until diminishing due to the gain in dominance of the Alpha variant in February 2021 [381].

- **B.1.1.207,** first reported in August 2020 in Nigeria, harbors P681H. As of late December 2020, this variant accounted for around 1% of viral genomes sequenced in Nigeria. As of May 2021, it had been detected in 10 countries.
- **B.1.1.318** (aka **VUI-21FEB-04** or **VUI-202102/04** by PHE on February 24, 2021, and belonging to GISAID clade GR and NextStrain clade 20B): 16 cases of it have been detected in the UK, and it is largely prevalent in Mauritius [382]. It is characterized by 14 non-synonymous mutations in the S gene, with 5 encoded amino acid substitutions (T95I, E484K, D614G, P681H, D796H) and ΔY144 in the Spike glycoprotein. It also harbors I82T in M and del309 in N. It was designated as Alert for further monitoring by WHO on June 2, 2021.
- **B.1.466.2** was designated by WHO an Alert for further monitoring on April 28, 2021, and was first reported from Indonesia
- **B.1.1.519** (belonging to GISAID clade GR and NextStrain clade 20 B) in Mexico harboring the mutations T478K, P681H, and T732A, which rapidly outcompeted the preexisting variants in 2021 [383]. It was designated as Alert for further monitoring by WHO on June 2, 2021.
- **B.1.616** is a VUI which caused a nosocomial outbreak in Western France in January 2021, poorly detected by RT PCR on nasopharyngeal samples, with high lethality (HR 4.2) [384]). It is characterized by 9 amino acid changes and one deletion in the S protein (H66D, G142V, Y144del, D215G, V483A, D614G, H655Y, G669S, Q949R, N1187D) in comparison with the original Wuhan strain, several unique amino acid changes in the E (F20L, T30I), M (H125Y), and N (T325I) proteins, in ORF1ab (T265I, N1324S, T1638I, S2261Y, Y3160H, L3606F, P314L, Q813R, L1681F, T2537I, K2674R) and ORF3 (Q57H) as well as by a deletion and frameshift in 2 proteins that antagonize various steps of type I interferon (IFN-I) production and signaling: ORF6 (del23-32 + frameshift resulting in 5 additional aa (HKPHN) at C-terminus) and replacement of the stop codon of ORF7a (*122R) resulting in a 5 amino acids extension at its C-terminus. It has not been detected after April 2021.
- **B.1.618** was first reported in October 2020. Spike has mutations Δ145-146, E484K and D614G, and is partly resistant to convalescent sera, bamlanivimab and mRNA vaccines-elicited antibodies [385]. As of April 23, 2021, the CoV-Lineages database showed 135 sequences detected in India, with single-figure numbers in each of 8 other countries worldwide
- **B.1.620**, harboring Spike mutations E484K, S477N and deletions HV69del, Y144del, and LLA241/243del, was imported from Central Africa to Europe [386].
- **B.1.621** (also named **21H** in NextStrain and belonging to GISAID GH clade) from Colombia in January 2021 with the insertion 145N in the N-terminal domain and amino acid change N501Y, E484K, and P681H [387], was designated an Alert for further monitoring by WHO on May 26, 2021.
- **C.36.3** and **C.36.3.1** (also known as **VUI-21MAY-02** by PHE, and belonging to NextStrain clade 20D and GISAID clade GR) was reported in multiple countries

since January 2021 [388]. It harbors A43S in ORF 7b, I82T in M, and R203K, G204R and G212V in N. It was designated as Alert for further monitoring by WHO on June 16, 2021.

- **R.1** lineage from Japan and Arizona (USA) harboring W152C, E484K, N501Y and G769V [193], which has become prevalent in Tokyo as of March 2021 but for which there is currently no evidence of increased severity [389]. Together with **R.2** lineage, it falls within GISAID GR lineage and NextStrain 20B and was designated by WHO as an Alert for further monitoring on April 7, 2021.

There is a significant and stepwise increase in RBD-ACE2 affinity at low temperatures, resulting in slower dissociation kinetics. This translated into enhanced interaction of the full Spike to ACE2 receptor and higher viral attachment at low temperatures [390]. Interestingly, variants harboring the N501Y mutation bypass this requirement and exhibited an increase in ACE2 binding compared to D614G Spike at both 4°C and 37°C, while this phenotype was only observed at 37°C with Spike from B.1.617.1, B.1.617.2, B.1.526 and B.1.429 [155].

According to a mathematical model and the speed of Spike evolution, another upcoming COVID-19 peak would come around July 2021 and disastrously attack Africa, Asia, Europe and North America [391].

Chapter 7
Characterization of SARS-CoV-2 Variants

Abstract Sequencing of the Spike gene is definitively the gold standard to define variants because it allows the recognition of all mutations/deletions/insertions affecting the whole viral sequence. However, this method is technically laborious and time-consuming.

Sequencing of the Spike gene is definitively the gold standard to define variants because it allows the recognition of all mutations/deletions/insertions affecting the whole viral sequence. However, this method is technically laborious and time-consuming. Currently, the protocol for SARS-CoV-2 whole genome sequencing developed by the ARTIC consortium (https://artic.network/ncov-2019) is applied worldwide but requires 4–5 days. A faster, but error-prone method is sequencing of the RBD segment. Nevertheless, sequencing cannot always be scaled or implemented in some settings. HiSpike is a novel three-step method for high-throughput targeted next generation sequencing (NGS) of the Spike gene in less than 30 h: it can sequence tenfold more samples compared to the conventional ARTIC method and at a fraction of the cost [392].

Alternative, easier-to-perform approaches including variant-specific RT-PCR and/or RFLP analysis of selected RT-PCR amplicons are awaiting full validation in the field. As previously explained, ΔHV69-70 causes SGTF in the Applied Biosystems TaqPath® COVID-19 PCR assay (Thermo Fisher), but under changed ecology, it is no longer conclusive for the B.1.1.7 variant. Nevertheless, in the US and other countries, screening samples for the SGTF helped to identify potential B.1.1.7 variants for sequencing prioritization. Thermo Fisher, however, has not released their Spike probe sequence, so the assay needs to be recreated to be used more broadly. Vogels et al. designed a ΔHV69-70 primer set that was able to distinguish between variant and non-variant samples similar to the TaqPath® SGTF signature. They combined the ΔHV 69/70 set in a multiplexed PCR assay with the CDC N1 set as a positive control, and the CDC RNase P set as an extraction/sample control as an open-source method to screen for viruses like B.1.1.7 with the ΔHV69/70 deletion. Essentially, it was an open-source "hack" of the TaqPath® assay [393]. Multiplex mutation-specific PCR-based assays with same-day reporting have also been developed:

D. Focosi, *SARS-CoV-2 Spike Protein Convergent Evolution*,
SpringerBriefs in Microbiology,
https://doi.org/10.1007/978-3-030-87324-0_7

- a PCR which differentially detects N501Y and ΔHV69-70 has been proposed an effective screening for samples worth of being sequenced, and able to discriminate potential B.1.1.7 (Δ69-70HV $^+$N501Y $^-$) from P.1 and B.1.351 (ΔHV69-70$^-$N501Y$^+$): the test was 100% specific when compared to PCR, with a limit of detection of 5000 copies/ml [394].
- another multiplex PCR was able to detect L450R, E484K, N501Y and ΔHV69-70 [395]

Finally, reverse-transcription reverse-complement polymerase chain reaction (RT-RC-PCR) has been proposed to rationalize reverse transcription and NGS library preparation into a single closed tube reaction [396].

Chapter 8
Predicting the Functional Consequences of Mutations

Mapping sequence data with the available structures from the Protein Data Bank (PDB), it is possible to generate hypothesis about the role of Spike mutations in ACE2 binding. Methodologically, there are several possible approaches. The most obvious is moving from patients getting reinfected from different clades, or from mutations detected in circulating Spike variants, and to verify neutralization from convalescent sera collected during the previous waves [18, 100].

In silico modeling can also be used. The ddG represents the difference in protein–protein affinity upon mutation: it can be measured using the Rosetta Flex ddG method and validated using surface plasmon resonance [397, 398]. GRID-based pharmacophore model (GBPM) has been used to identify mutations in both Spike and ACE2 that reciprocally affect binding [399]. Another computational model has been developed based on structure-dynamics-energy-based strategy [400]. All-atom steered molecular dynamics (SMD) simulations and microscale thermophoresis (MST) experiments have also been used to characterize the binding interactions between ACE2 and RBD [401].

As a third possibility, deep mutational scanning (DMS) predicts protein expression, ACE2 binding and mAb binding [142]. The method was first deployed with yeast display libraries [137] and then evolved to phage display libraries (https://jbloomlab.github.io/SARS-CoV-2-RBD_MAP_clinical_Abs/) [44] and finally mammalian cell surface display [402]. nAb binding is common within the fusion peptide and in the linker region before heptad repeat (HR) region 2. The complete escape maps forecast SARS-CoV-2 mutants emerging during treatment with mAbs and allow the design of escape-resistant nAb cocktails. DMS was also applied to polyclonal antibodies in CCP [403].

Lastly, mapping crystallographically determined interfaces between Spike mutants and nAb which do not disrupt ACE2 binding [404].

The "Genome to Phenotype (G2P)"-UK National Virology Consortium (https://www.ukri.org/news/national-consortium-to-study-threats-of-new-sars-cov-2-variants/) was the first newly created institution born to make such predictions.

© The Author(s), under exclusive license to Springer Nature Switzerland AG 2021 75
D. Focosi, *SARS-CoV-2 Spike Protein Convergent Evolution*,
SpringerBriefs in Microbiology,
https://doi.org/10.1007/978-3-030-87324-0_8

Chapter 9
Efficacy of Anti-Spike Vaccines and Monoclonal Antibodies Against SARS-CoV-2 Variants

Abstract This chapter summarizes available in vitro evidences of single vaccines and therapeutics against the main SARS-CoV-2 variants, and discusses features such as duration of protection, postponing second doses, heterologous boosting, third doses to immunosuppressed patients who did not seroconvert, and third doses to immunocompetent patients to counteract decline in antibody levels.

This paragraph represents a huge update of a review previously published in June 2021 from this author and colleagues [405]. Table 9.1 reports available evidences to date. Interpretation of results was simplified using a semiquantitative scale according to the number of folds decrease in neutralization efficacy. For each variant, the estimated reinfection rates and the proven reinfection cases (strains from each episode sequenced) are reported. Each variant was indicated using both the official (PANGOLIN and NextStrain) and the local (VUI/VOC/VOI) naming systems, and colloquial terms (e.g., "UK variant") in order to provide comprehensive association. The main, alarming finding is the lack of efficacy of single-agent bamlanivimab against most E484K-carrying variants. Accordingly, the FDA has recently withdrawn its emergency use authorization as a single agent, leaving the authorization only for usage in combination with etesevimab. Nevertheless, Q493R mutation, causing resistance to both mAbs, has been recently reported by our group [200]. High-frequency Spike mutations R346K/S, N439K, G446V, L455F, V483F/A, E484Q/V/A/G/D, F486L, F490L/V/S, Q493R and S494P/L might compromise some of mAbs in clinical trials [406]. A large, anonymized study evaluated 25 clinical-stage therapeutic antibodies for neutralization activity against 60 pseudoviruses bearing Spikes with single or multiple substitutions in several Spike domains, including the full set of substitutions in B.1.1.7 (Alpha), B.1.351 (Beta), P.1 (Gamma), B.1.429 (Epsilon), B.1.526 (Iota), A.23.1 and R.1 variants. 14 of 15 single antibodies were vulnerable to at least one RBD substitution, but most combination and polyclonal therapeutic antibodies remained potent. Key substitutions in the Spike protein of SARS-CoV-2 variants can predict resistance to mAbs, but other substitutions can modify the effects [407].

Table 9.1 Efficacy of vaccines and monoclonal antibodies against SARS-CoV-2 variants

Therapeutic\variant	Variants of concern (VOC)				Variants of interest (VOI)							Variants under monitoring	
	B.1.1.7	B.1.351	P.1	B.1.617.2	P.2	B.1.427 B.1.429	P.3	B.1.525	B.1.526 with E484K or S477N	B.1.617.1	C.37	B.1.258	B.1.1.2 98
PANGOLIN name	B.1.1.7	B.1.351	P.1	B.1.617.2	P.2	B.1.427 B.1.429	P.3	B.1.525	B.1.526 with E484K or S477N	B.1.617.1	C.37	B.1.258 Δ	B.1.1.2 98
NextStrain name	20I/S:501Y.V1	20H/S:501Y.V2	20J/s:501Y.V3	21A/S:478K	20B/S.484K	20C/S.452R	20B/S:265C	20A/S:484K	20C/S.484K	21A/S:154K	20D	–	–
PHE name	VOC-20DEC-01	VOC-20DEC-02	VOC-21JAN-02	VUI-21APR02	VUI-21JAN-01 VUI-202101/01	–	VUI-21MAR-02	VUI-21FEB-03	–	VUI-21APR-01	–	–	–
WHO name	alpha	beta	gamma	delta	zeta	epsilon	theta	eta	iota	kappa	lambda	–	–
GISAID name	GRY (formerly GR/501)	GH/501Y.V2	GR/501Y.V3	G/452R.V3	GR	GH/452R.V1	GR	G/484K.V3	GH	G/452R.V3	GR/452Q.V1	–	–

(continued)

(continued)

Table 9.1 (continued)

	VUI/VOC 2020 12/01, UK variant (Y.V1)	501Y.V2 VOC 202012/02	B.1.1.28.1 B.1.1.248 VOC 202101/02		B.1.1.28.2 B.1.1.28(E484K)	CAL.20C/ L452R, West Coast variant							Cluster V
local name	UK variant			Indian variant			–	Nigerian variant	–	Indian variant	–	–	–
Country of first detection	South-East England, UK	South Africa	Amazonas, Brazil	India	Rio de Janeiro, Brazil	Southern California, USA	Brazil	Nigeria	New York, USA	India	Peru	Czech Republirk c, Slovakia	Denmark
Convalescent plasma (sera) from previous waves	→ [100, 239, 262, 267, 268, 331, 502-508]	→→→ [160, 262, 267, 290, 294-297, 362, 502, 503, 506, 507, 509, 510]	→→→ [324, 331, 506], = [313]	= [511], → [355, 385, 507, 512], →→ [362, 509]	= [513], →→→ [339] (hamsters)	→→ [280, 510]	?	?	→→ [350, 514]	?	→→ [160, 362]	→ [141]	→ [506]
Hyperimmune serum (polyvalent immunoglobulins)	= [515]	→→→ [515]	→→→ [515]	?	?	= [515]	?	?	?	?	?	?	?

Table 9.1 (continued)

Estimated reinfection rate	0.7% [251]	?	6.4% [316]	?	?	?	?	?	?	?	?	?	?
Proven reinfection	4 cases [274–276]	5 cases [276, 301]	1 case [318]	2 cases [337, 338]	?	?	?	?	?	?	?	?	?
mAbs-	Eli Lilly (AbCellera/Junshi)	etesevimab (LyCoV016, CB6 or JS016 / LY3832479)	→↓↓ [131, 267]	= [267]	↓↓↓ [131, 324]	?	?	?	→↓ [280]	?	= [350]	?	?	?
		bamlanivimab (LY-CoV555 / LY3819253)	→↓↓ [264, 267, 290, 293]	= [264, 267]	↓↓↓ [290, 293, 324]	↓↓↓↓↓ [354, 355, 385, 512, 516]	↓↓↓ [264]	↓↓↓ [133] [280]	?	↓↓↓ [350, 514]	?	?	?
		LY-CoV1404	= [517]	= [517]	= [517]	?	= [517]	?	= [517]	?	?	?
	Regeneron	imdevim	= [267]	= [267, 362]	=	= [362]	= [280]	= [517]	= [350]	?	→	↓↓↓	= [518]

(continued)

Table 9.1 (continued)

on/Roche REGN-COV2 (Ronapreve®) ab (REGN10987)	↓↓↓ [131]	518]	[131, 324]				514]	[160, 362]	[141]	↓↓↓ [518]
casirivimab (REGN10933)	↓↓↓ [131]	↓↓↓ [267, 362, 518]	↓↓↓ [131, 290, 324]	↓↓↓ [362]	= [280]	?	↓↓↓ [350, 514]	= [160] [362]	?	↓↓↓ [518]
Celltrion regdanvimab (CT-P59)	↓↓ [519]	?	↓↓↓ [520]	↓↓↓ [521]	↓→ [280] ↓↓↓ [521]	?	↓↓↓ [521]	?	?	?
AstraZeneca AZD7442 /AZD8895/COV2-2196 long-acting antibody (LAAB) tixagevimab /AZD8895	↓↓↓ [267]	↓↓↓ [267]	= [324]	?	?	?	?	?	?	?
cilgavimab /AZD1061/COV2-2130	= [267]	= [267]	= [324]	?	?	?	?	?	?	?
BMS C135	= [267]	= [267]	?	?	?	?	?	?	?	?

(continued)

Table 9.1 (continued)

	C144	?	?	?	?	?	?	?	?	?	?
Vir Biotechnology	VIR-7831 / GSK-4182136 (sotrovimab) and VIR-7832 / GSK-4182137 (both derived from S309)	= [267, 522]	= [324, 522]	?	?	?	?	?	?	?	?
Menarini /TLS	MAD000 4J08	↓↓ [523, 524]	↓↓ [523, 524]	?	?	?	?	?	?	?	?
Vaccine BioNtech /Pfizer	BNT162b2/tozina meran (Comirnaty®) ↓ [259, 260, 262-264, 267, 290, 305, 331, 502, 504, 526-529] RBD	=/↓ [239, 257-264, 267, 502, 504, 526-528] ↓↓ [224, 362, 529] RBD [531] COVID19	=/↓↓ [304, 324, 325, 331, 504, 528] = [313]	↓↓ [224, 354-356, 362, 385, 512, 529, 535-537]	→ [264]	=/↓↓ [225, 280, 528] RBD [531]	→ [536]	↓ [225, 350, 514] ↓ [224]	↓↓ [537] → [536]	↓ [160] ↓↓ [224, 362]	?	= [528]

(continued)

Table 9.1 (continued)

AZD2816	?	= [546]	?	?	?	?	?	?	[534, 538, 545]	= [546]		
J&J / Janssen JNJ-7843673 5/Ad26.COV2.S	?	↓↓↓ [547] ↓↓ [548]	?	?	?	?	?	?	→ [548]	→ [548]		
Gamaleya Sputnik V/Gam-Covid-Vac ↓↓↓ [549]	= [549]		?	?	?	?	?	?	?			
Novavax NVX-CoV2373 (Covovax®) COVID19: 51% [550]	→ [539] ↓↓↓ [510]	↓ [539]	?	?	↓ [510]	?	?	?	?			
Bharat Biotech BBV152 / Covaxin ?	= [269] ↓ [509]	= [511] ↓ [509, 551] COVID19 65%	→ [513]	?	?	?	?	?	?			

(continued)

Table 9.1 (continued)

Moderna mRNA-1273 (Spikevax®)	↓↓↓ [304, 530, 531] COVID19 75% [532]-87%[533]-93.7% 93%[534]	90% [270, 532]-		COVID19 79-88% [534, 538]	?	=/↓↓ [280, 510, 528]	?	?	↓↓ [350, 514] → [224]	?	→ [160] ↓↓ [224, 362]	?
	↓ [266, 267, 295, 305, 377, 527, 528, 539] ↓ NHP COVID19 [540]	= [266-268] ↓ [527, 528, 541] ↓↓↓ [224, 362, 377, 510]	=/↓ [324, 528] ↓↓ [377]	→ [377] ↓↓ [224, 362, 385, 535, 537]								= [528]
AstraZeneca AZD1222 / ChAdOx1 (Vaxzevria®, Covishield®)	↓↓ [307] = hamster COVID19[542] COVID19 - 74.5%[534	↓↓[362] COVID19 90% [270]- = hamster COVID19 [542]	↓↓ [325]	↓ [512, 543, 544] ↓↓↓ [362] COVID19 60-67%	?	?	?	?	?	?	↓↓↓ [362] ?

(continued)

Table 9.1 (continued)

SinoVac	CoronaVac			[552]					
↓↓↓ [553] ↓↓[507, 554]	= [507, 553]	↓↓ [553] ↓[554]	↓ [555] [507]	?	= [553]	?	?	↓→ [553]	↓→ [554] ?

Arrows indicate fold-reductions in neutralizing activity compared to wild-type D614G strain (e.g.: Wuhan-Hu-1, USA-WA1/2020, B.1, or other reference strains) (= no reduction; ↓ : 1-3 fold reduction; ↓↓ : 3-5 fold reduction; ↓↓↓ : > 5 fold reduction; ? : data not available).

COVID19 : vaccine efficacy against symptomatic diseases in humans (if not otherwise indicated) or in animal models (specified).

NHP: nonhuman primates.

RBD: ACE2-RBD competition assay.

VOC: variants of concern

VOI : variant of interest

While the in vitro findings summarized here wait for confirmatory clinical evidences, in the meanwhile, they could orient therapeutic and preventive strategies.

No in vitro evidences of efficacy against SARS-CoV-2 variants have been published for several widely used vaccines (e.g., Sinopharm's BBIBP-CorV) or only predictions were available in other cases (e.g., CureVac's CVnCOV [408]), stressing the need for more studies.

Meta-analysis of 56 vaccine studies, including 2,483 individuals and 8,590 neutralization tests, showed that, compared with lineage B, there was a 1.5-fold reduction in neutralization against the B.1.1.7, 8.7-fold reduction against B.1.351 and 5.0-fold reduction against P.1. The estimated neutralization reductions for B.1.351 compared to lineage B were 240.2-fold reduction for non-replicating vector platform, 4.6-fold reduction for RNA platform and 1.6-fold reduction for protein subunit platform. The nAbs induced by administration of inactivated vaccines and mRNA vaccines against lineage P.1 were also remarkably reduced by an average of 5.9-fold and 1.5-fold [409]. Efficacy of CCP from previous waves is also generally lowered: CCP from a donor affected during the early 2020 protected against SARS-CoV-2 WA-1 wild-type strain but was insufficient to protect against challenge with B.1.1.7 and B.1.351 in a mouse model [410].

The **correlation between these in vitro data and real-life vaccine efficacy (VE)** shows that nAb levels are a good correlated of protection (CoP). Khoury et al., assuming that the neutralization level required for 50% protection against detectable SARS-CoV-2 infection was 20.2% of the mean convalescent sera level and that the neutralization level for 50% protection from severe infection was 3% of the mean convalescent sera level, reported that the decay of the neutralization titer over the first 250 days after immunization predicts a significant loss in protection from SARS-CoV-2 infection, although protection from severe disease should be largely retained. Neutralization titers against some SARS-CoV-2 VOC were reduced compared with the vaccine strain [411]. Similarly, Earle et al. evaluated the relationship between efficacy and in vitro neutralizing and binding antibodies of 7 vaccines. Once calibrated to titers of human convalescent sera reported in each study, a robust correlation was seen between neutralizing titer and efficacy ($\rho = 0.79$) and binding antibody titer and efficacy ($\rho = 0.93$), despite geographically diverse study populations subject to different forces of infection and circulating variants, and use of different endpoints, assays, convalescent sera panels and manufacturing platforms [412]. Accordingly, breakthrough infections among 1497 BNT162b2-vaccinated healthcare workers happened in those with lower nAb titers during the peri-infection period (case-to-control ratio, 0.361), and higher peri-infection neutralizing antibody titers were associated with lower infectivity (higher Ct values). Similarly, mRNA-1273-elicited sera binding and nAb titers correlated with COVID-19 risk and vaccine efficacy and likely have utility in predicting mRNA-1273 vaccine efficacy against COVID-19. Modelling frameworks can identify correlates of protection based on live SARS-CoV-2 variants nAb titres from vaccinated individuals. They have been to predict vaccine effectiveness in overall populations and age subgroups. It was validated by predicting effectiveness against the B.1.167.2 (Delta) variant. The predictions, of 51.7% (34%, 69%) after one and of 88.6% (76%, 97%) after two vaccine doses, were close to the corresponding means, 49% and 85.4%, of observations in real-life effectiveness studies.

A case–control study showed that, compared with unvaccinated individuals, BNT162b2 vaccinees with documented SARS-CoV-2 infection at least a week after the second dose were disproportionally infected with B.1.351 (odds ratio of 8:1), while those infected between 2 weeks after the first dose and 1 week after the second dose, were disproportionally infected by B.1.1.7 (odds ratio of 26:10), suggesting reduced VE against both VOCs under different dosage/timing conditions. Nevertheless, the B.1.351 incidence in Israel to date remains low and vaccine effectiveness remains high against B.1.1.7, among those fully vaccinated. These results overall suggest that vaccine breakthrough infection is more frequent with both VOCs, yet a combination of mass-vaccination with two doses coupled with non-pharmaceutical interventions control and contain their spread [413]. With mRNA-1273, binding of vaccine-elicited anti-RBD antibodies is more broadly distributed across epitopes than for infection-elicited anti-RBD antibodies [414]: accordingly, greater IgG_2, IgG_3 and IgG_4 responses and higher ratios of $(IgG_1 + IgG_3)/(IgG_2 + IgG_4)$ were seen in subjects vaccinated with either BNT162b2 or mRNA-1273 than in convalescents [415]. This greater binding breadth means single RBD mutations have less impact on neutralization by vaccine sera than convalescent sera. Another striking feature is that for BNT162b2 post-first dose vaccination infection did not increase IgG titers, so that individuals infected post-dose one should receive the second [416]. CoronaVac for which no in vitro efficacy data are available was 42% effective in the real-world setting of extensive P.1 transmission, but significant protection was not observed until completion of the two-dose regimen [417]. A meta-analysis by Shapiro et al. found that, on average, the vaccine efficacy (VE) against any disease with infection was 85% after a full course of vaccination. The VE against severe disease, hospitalization or death averages close to 100%. The average VE against infection, regardless of symptoms, was 84%. The average VE against B.1.1.7 [418] B.1.1.28 (P1) and B.1.351 was 86%, 61% and 56%, respectively [419]. Questions also stem about the clinical value of different overall antibody levels after each type of vaccine: possibly due to the higher S-protein delivery, titers peaked at week 6 for both vaccines, but were significantly higher for mRNA-1273 vaccine on days 14–20 ($p < 0.05$), 42–48 ($p < 0.01$), 70–76 ($p < 0.05$), 77–83 ($p < 0.05$), and higher for BNT162b2 vaccine on days 28–34 [420].

Additionally, a single injection of mRNA-1273 or BNT162b2 has been shown enough to induce **novel antibody specificities** that protect against the B.1.351 VOC [305, 421]: a similar phenomenon has been reported after 2 BNT162b2 doses against B.1.1.7 [422]. NAbs titers increased in previously infected BNT162b2 vaccinees relative to uninfected vaccinees against every variant tested: 5.2-fold against B.1.1.7, 6.5-fold against B.1.351, 4.3-fold against P.1 and 3.4-fold against original SARS-CoV-2 [423]. Similarly, a single dose of either BNT162b2 or ChAdOx1 vaccines in convalescents raised the titer of antibodies against the SARS-CoV-2 vaccine strain (B.1) and three major VOCs (B.1.1.7, B.1351 and P.1). Lineages with E484K and N501Y/T (e.g., B.1.351 and P.1) have the greatest reduction in neutralization, followed by lineages with L452R (e.g., B.1.617.2) or with E484K (without N501Y/T). While both groups retained neutralization capacity against all variants, plasma from previously infected vaccinated individuals displayed overall better neutralization capacity when compared to plasma from uninfected individuals that also received two vaccine doses [424]. Vaccination and natural SARS-CoV-2 infection elicit neutralizing antibody responses that are most potent against variants that bear spike mutations present in the immunizing exposure. This trend is exemplified by variants with mutations at the spike E484 position, which were all neutralized more effectively by E484K-exposed serum than other serum types. Importantly, it has been shownthat B.1.617.2 (Delta) is neutralized more effectively by serum elicited by prior exposure to two different variants — B.1.429 and B.1.1.519 — which have separate subsets of spike mutations overlapping with mutations in B.1.617.2. Given that different regions throughout the world have experienced variable transmission of different variants prior to the dominance of B.1.617.2, these results suggest that acquired immunity in the population will differ significantly depending on the previous prevalence of variants in each region. Furthermore, these results demonstrate that specificity is strongest for serum neutralizing variants fully homologous to the exposure, but even single shared spike mutations, particularly those in highly antigenic regions such as the RBD, can enhance cross-neutralization.

A **single vaccine dose to convalescents** is nowadays a well-accepted approach that saves money and side effects [425]: e.g., one dose of the BNT162b2 vaccine increases nAb titers against the B.1.1.7, B.1.351 and P.1 variants in persons previously infected with SARS-CoV-2 [426].

Apart from efficacy, many topics remain under investigation for anti-Spike vaccines:

- **vaccine-elicited T cell immunity**: while nAbs are just one arm of the adaptive immune response to vaccines, very few data are available for protection from T cell immunity, which would be especially relevant in the ones who do not mount antibody responses. Gallagher et al. found detectable but diminished T cell responses to Spike variants (B.1.1.7, B.1.351 and B.1.1.248) among BNT162b2 or mRNA-1273 vaccinated donors [427]. BNT162b2 or mRNA-1273-elicited Spike-specific T cells responded similarly to stimulation by Spike epitopes from the ancestral, B.1.1.7 and B.1.351 variant strains, both in terms of cell numbers and phenotypes. In infection-*naive* individuals, the second dose boosted the quantity but not quality of the T cell response, while in convalescents, the second dose helped neither. Spike-specific T cells from convalescent vaccinees differed strikingly from those of infection-naive vaccinees, with phenotypic features suggesting superior long-term persistence and ability to home to the respiratory tract including the nasopharynx [428].
- **duration of protection**: according to a mathematical model by Luo et al., after mRNA-1273 vaccination, pseudovirus neutralization test against B.1.351 is expected to fall below the lower limit of detection of 20 geometric mean titers on day 100; variant P.1 on day 202, variant B.1.429 on day 258; and variant B.1.1.7 on day 309 [429]. Real-world data instead suggested that binding and functional antibodies against B.1.1.7, B.1.351, P.1, B.1.429 and B.1.526 variants persisted in

most subjects, albeit at low levels, for 6 months after the primary series of mRNA-1273 [430]. A similar declining trend in VE against symptomatic COVID-19 (−6% every 2 months) was proven in real life at 6 months for BNT162b2 [431]. A model predicts and exemplifies several possible consequences for vaccine efficacy in VOC infections: (1) a delay in the onset of vaccine efficacy against VOC; (2) a transient increase in susceptibility to breakthrough infection with VOC compared to non-VOC as a function of time after vaccination. Preliminary data indicate that such phenomena are observed in studies of the B.1.1.7 and B.1.351 variants. Ignoring the strong dependence on the time post-vaccination can lead to contradictory reports of relative efficacy against VOC versus non-VOC, with implications on mitigation strategies against VOC and the design of vaccine efficacy studies [432]. nAb titres declined substantially six months after two doses of CoronaVac among older adults, but a booster dose rapidly induces robust immune responses. Natural immunity confers longer lasting and stronger protection against infection (RR = 5.96), symptomatic disease (RR = 7.33) and hospitalization caused by the Delta variant of SARS-CoV-2, compared to the BNT162b2 two-dose vaccine-induced immunity. Individuals who were both previously infected with SARS-CoV-2 and given a single dose of the vaccine gained additional protection against the Delta variant. In conclusion, BNT162b2-induced protection against infection appears to wane rapidly after its peak right after the second dose, but it persists at a robust level against hospitalization and death for at least 6 months following the second dose. At 6-months after first BNT162b2 dose, compared to HCW, elderlies have significantly lower anti-SARS-CoV-2 S1-, full Spike- and RBD-IgG seropositivity rates, IgG levels, serum neutralization of Delta VOC and T cell reactivity. In July 2021 in Israel, the rates of both documented SARS-CoV-2 infections and severe COVID-19 exhibit a statistically significant increase as time from second vaccine dose elapsed. Elderly individuals (60+) who received their second dose in March 2021 were 1.6 times more protected against infection and 1.7 times more protected against severe COVID-19 compared to those who received their second dose in January 2021. Similar results were found for different age groups.

- **postponing second doses** has been widely implemented in order to optimize vaccine delivery under manufacturing bottlenecks. In non-convalescent elderlies higher than age 80 who received the second dose of BNT162b2 after 12 weeks instead of 3, the peak antibody response was 3.5-fold higher, but cellular immune responses were 3.6-fold lower [433].
- **heterologous boosting**: heterologous immunization strategy combining inactivated and mRNA vaccines can generate robust vaccine responses and therefore provide a rational and effective vaccination regimen [434]. ChAdOx/BNT162b2 booster vaccination was largely comparable to homologous BNT162b2/BNT162b2 vaccination and overall well tolerated [435]. No major differences were observed in the frequency or severity of local reactions after either of the vaccinations. In contrast, notable differences between the regimens were observed for systemic reactions, which were most frequent after prime immunization with ChAdOx (86%) and less frequent after homologous BNT162b2/BNT162b2 (65%), or heterologous ChAdOx/BNT162b2 boosters

(48%) [436–438]. Among 229 vaccinees that received a BNT162b2 boost 9 to 12 weeks after ChAdOx1 nCoV-19 prime, nAb titers were significantly higher than after homologous ChAdOx1 nCoV-19 and even than homologous BNT162b2 vaccination [439]. Heterologous regimen induced Spike-specific IgG, nAbs, and Spike-specific CD4[+] T cells, which were significantly more pronounced than after homologous vector boost, and higher or comparable in magnitude to the homologous mRNA regimens. Moreover, Spike-specific CD8 T cell levels after heterologous vaccination were significantly higher than after both homologous regimens [440]. Neutralizing activity against the prevalent strain B.1.1.7 was 3.9-fold higher than in individuals receiving homologous BNT162b2 vaccination, only twofold reduced for variant of concern B.1.351, and similar for variant B.1.617 [441]. While both ChAdOx and BNT162b2 boosted prime-induced immunity, BNT162b2 induced significantly higher frequencies of Spike-specific CD4 and CD8 T cells and, in particular, high titers of nAbs against the B.1.1.7, B.1.351 [442] and the P.1 VOCs [443]. Heterologous ChAdOx1 followed by BNT162b2 induces moderate responses against alpha and delta, but poor responses against beta and gamma VOC. When CoronaVac vaccinees were boosted with either BNT162b2 or ChAdOx1 nCoV-19, increase in anti-S-RBD antibodies and surrogate neutralizing antibodies against the Delta variant was higher in BNT162b2 recipients than in ChAdOx1 nCoV-19 recipients.

- third dose to immunosuppressed patients who do not mount protective response after 2 doses: this approach benefits (both in nAb levels and Spike-reactive B and CD4+ T lymphocytes) a fraction of solid organ transplant recipients who did not respond to 2 doses of BNT162b2 (with the third dose being either homologous or heterologous ChAdOx1) [495] or 2 doses of mRNA-1273 [496], or thoracic cancer patients who did not respond to 2 doses of BNT162b2 [497]. Heterologous third dose with ChAdOx after 2 doses of BNT162b2 or mRNA-1273 to nonresponders treated with rituximab induces levels similar to a third homologous dose.

- third dose to immunocompetent subjects to counteract decline in serological response. With the evidence of waning immunity of the BNT162b2 vaccine, a national third dose vaccination campaign was initiated in Israel during August 2021; other countries have announced their intention to administer a booster shot as well. a third dose also boosts declining nAb titers in immunocompetent patients, which neutralize most VOCs/VOIs. Patients with the higher response rate to the third dose of vaccine can be identified by the presence of low anti-RBD IgG titers and Spike-specific CD4+ T cells in their circulation 14 days after the second dose. E.g., a third BNT162b2 dose restores the drop in viral load of Delta which otherwise would vanish after 6 months from second dose. Since 12 days after the third dose of BNT162b2, an 11.4-fold decrease in the relative risk of confirmed infection was seen, and a >10-fold decrease in the relative risk of severe illness. Another study similarly found that 7–13 days after the third shot there is a 48–68% reduction in the odds of testing positive for SARS-CoV-2 infection and that 14–20 days after the booster the marginal effectiveness increases to 70–84%. Although individuals fully vaccinated with Vaxzevria (AstraZeneca) have higher antibody levels than those with CoronaVac (Sinovac), heterologous prime-boost with CoronaVac-Vaxzevria yielded comparable antibody levels to two-dose Vaxzevria. On the reverse, participants who received the booster of AZD1222 possessed higher levels of spike RBD-specific IgG, total immunoglobulins, and anti-S1 IgA than that two-dose CoronaVac vaccines ($p < 0.001$) : they also elicited higher neutralizing activity against the wild type and all variants of concern than those in the recipients of the two-dose vaccines. Similarly, a third dose of BBIBP-CorV boosted humoral and cellular responses, even in those seronegatives after 2 doses.

- **delivery routes**: routes other than intramuscular lead to dose sparing, e.g., intradermal administration of 10 and 20 μg mRNA-1273 vaccine was well tolerated and safe, and resulted in a robust antibody response [444]. Current generation of intramuscularly delivered vaccines induces poor IgA in mucosae, with levels declining after the second dose after BNT162b2 or mRNA-1273 [445]. Since sterilizing immunity is required to stop transmission and achieve herd immunity, mucosal vaccines are being investigated. Accordingly, VOC delta viral loads are the same in vaccinated and non-vaccinated cases [446]. While the *in vitro* findings summarized here wait for confirmatory clinical evidences, in the meanwhile they could orient therapeutic and preventive strategies.

Chapter 10
Selective Pressures Exerted by Antibody-Based Therapeutics

Evolutionary modeling suggests that SARS-CoV-2 strains harboring 1–2 deleterious mutations naturally exist, and their frequency increases steeply under positive selection by mAbs and vaccines [447]. In 2% of COVID cases, viral variants with multiple mutations, including in the Spike glycoprotein, can become the dominant strains in as little as one month of persistent in-patient virus replication. This suggests the continued local emergence of VOC independent of travel patterns [448].

While mutations can occur as a consequence of small chemicals (e.g., molnupiravir [449]), it is reasonable that widespread deployment of nAb-based therapeutics could accelerate Spike immune escape. As previously explained, DMS maps identify mutants arising after treatment with REGN-COV2: it is of interest that such escape mutants already circulate [44].

In vitro, continuous passaging of SARS-CoV-2 in the presence of a CCP unit with nAb titer $> 1:10^4$ led to ΔF140 at day 45, followed by E484K at day 73, and an insertion in the NTD: these accumulating mutations led to complete immune escape [450]. Accordingly, K417N, E484K and N501Y mutations were selected when pseudotyped SARS-CoV-2 was cultured in the presence of vaccine-elicited mAbs [139]. In vivo, while intra-host SARS-CoV-2 mutation development is typically very low [451], faster MR have been found in longitudinal studies of immunodeficient patients who had persistent SARS-CoV-2 infections for several months and were treated with nAb-based therapeutics:

- anti-Spike mAbs:

  - REGN-CoV2 cocktail:

    Choi et al. reported a case having detectable SARS-CoV-2 for 154 days, with accelerated viral evolution in the Spike protein after treatment with remdesivir and the anti-Spike REGN-CoV2 mAb cocktail [452]. Hamster models and clinical trials showed no emergence of variants [138].

D. Focosi, *SARS-CoV-2 Spike Protein Convergent Evolution*, SpringerBriefs in Microbiology, https://doi.org/10.1007/978-3-030-87324-0_10

- AZD7442 (COV2-2130 and COV2-2196) cocktail was resistant to rapid escape [453].
- bamlanivimab monotherapy:

  Truffot et al. reported emergence of E484K and Q493R after treatment with bamlanivimab [202]

  Lohr et al. reported rapid selection of immune escape variant carrying E484K mutation in a B.1.1.7 infected and immunosuppressed patient treated with bamlanivimab [454].

- bamlanivimab + etesevimab cocktail:

  Focosi et al. reported one immunocompromised patient with cholan-giocarcinoma under steroids developing Q493R after treatment with bamlanivimab-etesevimab who finally died [455].

  Guigon et al. reported each one immunocompromised patient with mycosis fungoides developing Q493R after treatment with bamlanivimab-etesevimab who finally recovered [201].

  Vellas et al. reported five immunocompromised patients: three of these patients harbored variants with a Q493R (detected on day 7 in two patients and on day 14 in the third). A Q493K mutation variant was detected in one patient on day 7 post-treatment, and a E484K mutation variant was found in another on day 21.

- CCP: Immune escape under CCP has not been reported very commonly nor fastly. E.g., none out of eight recipients of hematopoietic stem cell transplants or chimeric antigen receptor T lymphocytes treated with CCP and testing SARS-CoV-2 positive for 2 months showed significant mutations compared to the original strain [456].

  - Avanzato et al. reported within-host genomic evolution in a patient affected by chronic lymphocytic leukemia and iatrogenic hypogammaglobulinemia who received CCP and shed infectious SARS-CoV-2 for 70 days [457].
  - Hensley et al. reported a CAR-T-cell recipient developed severe COVID-19, intractable RNAemia and viral replication lasting > 2 months while receiving remdesivir and low-titer non-neutralizing CCP (day 2 and 58) and developed multiple variants [458].
  - Kemp et al. reported an immune suppressed individual who developed D796H and ΔH69/ΔV70 mutations after each unsuccessful course of CCP. In vitro, such mutant showed similar infectivity to wild type strain but resistance to different CCP donors [459].
  - Truong et al. reported the emergence of seven major and three minor allele variants (including Δ141–143, Δ145, Δ141–144, Δ211–212, N440K, V483A and E484Q) in a patient with acute lymphoblastic leukemia who was treated with weekly CCP and tested persistently positive for SARS-CoV-2 until day 144 [198].

- Chen et al. documented the microevolution of SARS-CoV-2 recovered from sequential tracheal aspirates from an immunosuppressed patient on tacrolimus, steroids and CCP therapy and identify the emergence of multiple NTD and RBD mutations associated with reduced antibody neutralization as early as 3 weeks after infection. Comparison of SARS-CoV-2 genomes from the first swab (Day 0) and 3 tracheal aspirates (Day 7, 21 and 27) identified five different S protein mutations at the NTD or RBD regions from the second tracheal aspirate sample (21 Day). The S:Q493R substitution and S:243-244LA deletion had ~ 70% frequency, while ORF1a:A138T, S:141-144LGVY deletion, S:E484K and S:Q493K substitutions demonstrated ~ 30%, ~ 30%, ~ 20% and ~ 10% mutation frequency, respectively. However, the third tracheal aspirate sample collected one week later (Day 27) was predominated by the haplotype of ORF1a:A138T, S:141-144LGVY deletion and S:E484K (> 95% mutation frequency). Notably, S protein deletions (141-144LGVY and 243-244LA deletions in NTD region) and substitutions (Q493K/R and E484K in the RBD region) previously showed reduced susceptibly to mAb or CCP [460].
- Monrad et al. reported SARS-CoV-2-positive NPS persisting beyond 333 days in an immunocompromised patient with B-CLL, asymptomatically carrying infectious SARS-CoV-2 at day 197 post-diagnosis. In addition, viral sequencing indicates major changes in the Spike protein over time, temporally associated with CCP treatment, including H49Y, delY144, delLLA241-243, delAL243-244, L242H, A243P, F490S, N1178N and C1250F [461].
- Khatamzas et al reported a follicular lymphoma patient treated with remdesivir and CCP who developed L18F and R682Q.

In the absence of nAb-based therapeutics, immunosuppressive treatment has been rarely associated with Spike mutations [456]:

- Bazykin et al. reported emergence of Y453F and Δ69-70HV mutations ("the ΔF combination") (together with S50L, Δ141–144, T470N and D737G) in a 47-years old female with diffuse large B cell lymphoma treated with rituximab plus chemotherapy (R-ICE regimen) [462].
- Borges et al. reported another DLBCL patients with persistent infection for 6 months who developed four mutations (V3G, S50L, N87S and A222V) and two deletions (Δ18-30 and Δ141-144) in Spike [463].
- Truong et al. reported the emergence of escape mutations in two more patients with acute lymphoblastic leukemia who were persistently positive for SARS-CoV-2 for up to 162 days [198].
- Karim et al. reported evolution of E484K and N501Y mutations in a case of prolonged infection of greater than 6 months with the shedding of high titter SARS-CoV-2 in an individual with advanced HIV and antiretroviral treatment failure [464].
- Kavanagh Williamson et al. reported an hypogammaglobulinemic individual who was persistently infected with SARS-CoV-2 for over 290 days, the longest persistent infection recorded in the literature to date. During this time, nine samples

of viral nucleic acid were obtained and analyzed by next-generation sequencing. Initially, only a single mutation (L179I) was detected in the Spike protein relative to the prototypic SARS-CoV-2 Wuhan-Hu-1 isolate, with no further changes identified at day 58. However, by day 155, the Spike protein had acquired a further four amino acid changes, namely S255F, S477N, H655Y and D1620A, and a two amino acid deletion (ΔH69/ΔV70). Infectious virus was cultured from a nasopharyngeal sample taken on day 155, and NGS confirmed that the mutations in the virus mirrored those identified by sequencing of the corresponding swab sample. The isolated virus was susceptible to remdesivir in vitro; however, a 17-day course of remdesivir started on day 213 had no effect on the viral RT-PCR cycle threshold (Ct) value. On day 265, the patient was treated with the combination of casirivimab and imdevimab. The patient experienced progressive resolution of all symptoms over the next 8 weeks, and by day 311, the virus was no longer detectable by RT-PCR [465].

- Mendes-Correa et al. reported a persistent SARS-CoV-2 infection of at least 218 days in a male who had undergone a prior autologous hematopoietic stem cell transplant due to a diffuse large B cell lymphoma. He did not manifest a humoral immune response to the virus. Whole-genome sequencing and viral cultures confirmed a continual infection with a replication-positive virus that had undergone genetic variation for at least 196 days following symptom onset [466].
- Sepulcri et al reported a NHL male patient treated with remdesivir and CCP who developed H69Y-Pl, V70G and S982A at day 238.
- Weigang et al reported a transplant recipient treated with remdesivir who developed S13I, T95I, E484G, F490L, D141-144, D244-247, and D680-687.
- Lee et al reported intraclonal evolution in 18 B-cell non-Hodgkin lymphoma patients, associated with impaired CD8+ T-cell counts.Remdesivir can adopt both amino and imino tautomeric conformations to base-pair with RNA bases.

Both amino-remdesivir: G and imino-remdesivir: C pairs could be quite mutagenic. Serial in vitro passages of SARS-CoV-2Engl2 in cell culture media supplemented with remdesivir selected for drug-resistant viral populations. Remdesivir triggers the selection of SARS-CoV-2 variant with a E802D mutation in the RdRp sufficient to confer decreased sensitivity to remdesivir without affecting viral fitness. Another mutation, I168T, was observed in the Nsp6. The analysis of more than 200,000 sequences also revealed the occurrence of 22 mutations in the spike, including changes in amino acids E484 and N501 corresponding to mutations identified in alpha and beta. It has been hence been proposed than nAb-based therapeutics could amplify mutations induced by remdesivir.

Chapter 11
Which Strain Will Finally Become Dominant?

Abstract Vaccine campaigns are likely to affect viral evolution.

The answer lies in the respective vaccine efficacy (VE) against variants. Right now, we have some insights from theoretical models accounting for transmissibility and immune escape. Preliminary evidences suggest B.1.351 could be fitter than B.1.1.7 [467]. Yang et al. estimate that B.1.1.7 has a 46.6% increase in transmissibility but nominal immune escape from protection induced by prior wild-type infection; B.1.351 has a 32.4% increase in transmissibility and 61.3% immune escape; and P.1 has a 43.3% increase in transmissibility and 52.5% immune escape. Model simulations indicate that B.1.351 and P.1 could supplant B.1.1.7 dominance and lead to increased infections [468]. The same could be true for B.1.617.2 and B.1.1.318. In the USA, the percentage of SARS-CoV-2 positive cases that are B.1.1.7 dropped from 70% in April 2021 to 42% in just 6 weeks: rapid growth rates of variants B.1.617.2 (0.61) and P.1 (0.22) was the primary drivers for this displacement, with B.1.617.2 growing faster in counties with a lower vaccination rate [469]. Preliminary modeling by WHO based on sequences submitted to GISAID suggests that B.1.617 has a higher growth rate than other circulating variants in India, suggesting potential increased transmissibility [470, 471]: the same has been shown for B.1.617.2 over B.1.1.7 in UK [472], with doubling time between 5–14 days [473]. The frequency of the delta is expected to take over the alpha in Japan around July 12, 2021, at the time of the Olympic games [474]. In the absence of vaccine, the main driver in the evolutionary game is the efficacy of exposure from previous infection at preventing reinfection. Beta virus showed moderate (7-fold) and delta slight escape from neutralizing immunity elicited by ancestral virus infection. In contrast, delta virus had stronger escape from beta elicited immunity (12-fold), and beta virus even stronger escape from delta immunity (34-fold). The SARS-CoV-2 evolved within an HIV-1 patient had 9-fold escape from ancestral immunity, 27-fold escape from delta immunity, but was effectively neutralized by beta immunity. Beta and delta are serologically distant, further than each is from ancestral [558]. Infectivity (tipically defined as the ratio between infectious viral titer calculated with a focus forming assay and the mRNA copies of several genes) is another driver. In a systematic review, out of 276 positive-culture of non-severe patients, 272 (98.55%) were negative 10 days after symptoms onset, while PCR assays remained positive for up to 67 days. In severely ill or immunocompromised patients positive-culture was obtained up to 32 days and out of 168 cultures, 31 (18.45%) stayed positive after day 10. In non-severe patients, in Ct value greater than 30 only 10.8% were still culture-positive while in Ct >35 it was nearly universally negative. The minimal calculated number of viral genome copies in culture-positive sample was 2.5×10^3 copies / mL. These findings were similar in immunocompromised patients. Recovering positive culture from non-respiratory samples was sporadically obtained in stool or urine samples. Conversion of Ct values to viral genome copies showed variability between different PCR assays and highlighted the need to standardize reports to correctly compare results obtained in different laboratories. By aggregating VOC-associated and antibody-selected spike substitutions into a single polymutant spike protein, Schmidt et al showed that 20 naturally occurring mutations in SARS-CoV-2 Spike are sufficient to generate pseudotypes with near-complete resistance to the polyclonal neutralizing antibodies generated by convalescents or mRNA vaccine recipients. Strikingly, however, plasma from individuals who had been infected and subsequently received mRNA vaccination, neutralized pseudotypes bearing this highly resistant SARS-CoV-2 polymutant spike, or diverse sarbecovirus spike proteins.In the absence of vaccine, the maindriver in the evolutionary game is the efficacy of exposure from previous infection at preventing reinfection. Beta virus showed moderate (7-fold) and delta slight escape from neutralizing immunity elicited by ancestral virus infection. In contrast, delta virus had stronger escape from beta elicited immunity (12-fold), and beta virus even stronger escape from delta immunity (34-fold).The SARS-CoV-2 evolved within an HIV-1 patienthad 9-fold escape from ancestral immunity, 27-fold escape from delta immunity, but was effectively neutralized by beta immunity.Beta and delta are serologically distant, further than each is from ancestral.

Infectivity (typically defined as the ratio between infectious viral titer calculated with a focus forming assay and the mRNA copies of several genes) is another driver. In a systematic review, out of 276 positive-culture of non-severe patients, 272 (98.55%) were negative10days after symptoms onset, while PCR assays remained positive for up to 67 days. In severely ill or immunocompromised patients positive-culture was obtained up to 32 days and out of 168 cultures, 31 (18.45%) stayed positive after day 10. In non-severe patients, in Ct value greater than 30 only 10.8% were still culture-positive while in Ct >35 it was nearly universally negative. The minimal calculated number of viral genome copies in culture-positive sample was 2.5×10^3 copies /mL.These findings were similar in immunocompromised patients. Recovering positive culture from non-respiratory samples was sporadically obtained in stool or urine samples. Conversion of Ct values to viral genome copies showed variability between different PCR assays and highlighted the need to standardize reports to correctly compare results obtained in different laboratories.

By aggregating VOC-associated and antibody-selected spike substitutions into a single polymutant Spike protein, Schmidt et al. showed that 20 naturally occurring mutations in SARS-CoV-2Spike are sufficient to generate pseudotypes with near-complete resistance to the polyclonal nAbs generated by convalescents or mRNA vaccine recipients. Strikingly, however, plasma from individuals who had been infected and subsequently received mRNA vaccination, neutralized pseudotypes bearing this highly resistant SARS-CoV-2 polymutant Spike, or diverse sarbecovirus spike proteins.

Chapter 12
Conclusions

Abstract Much uncertainty remains about the dynamics of viral evolution. Fast development of vaccines and therapeutics will be required to counterfeat novel viral variants.

Current SARS-CoV-2 diversity and MR (1–2 SNPs per month [475]) is far lower than that seen for influenza viruses, but the pandemic is increasing genetic variation. National lockdowns have created a landscape where in-country evolution can be detected only after reopening of borders and travel to developed countries sequencing a lot of positive samples.

Virtually all anti-SARS-CoV-2 CD8$^+$ T-cell responses should recognize the recent variants [476], but antibody neutralization, the prerequisite for reaching herd immunity, will likely be impaired. This is not unexpected for a coronavirus: For example, the common cold coronavirus HCoV-229E evolved antigenic variants that are comparatively resistant to the older sera [477].

It will be better to use vaccines targeting the faster spreading SARS-CoV-2 strain, even when the initial prevalence of this variant is much lower [478]. Given the reported reduced neutralization by vaccine-elicited antibodies against single to triple K417N + E484K + N501Y mutants [139] and the dominating delta variant, vaccines will need to be updated periodically to avoid potential loss of clinical efficacy, and in this regard, mRNA vaccines are likely the easiest to be remanufactured.

mAb cocktails theoretically have the potential to minimize immune escape: While escape occurs when combining mAbs targeting overlapping regions of Spike, this does not happen when combining non-competing antibodies [479]. Nevertheless, novel mutants rapidly appear after treatment with individual mAb, causing loss of neutralization. Assuming that Spike affinity to ACE2 should be preserved in variants, ACE2-Ig proteins could be an effective weapon against SARS-CoV-2 variants [184].

CCP is likely to remain the fastest deployable weapon against a clinically significant viral variant [480]: it has been formally proven that only a minority of CCP samples lose all neutralizing activity in contrast to mAbs from five different epitope clusters, where neutralization was completely abrogated by a single Spike mutation. While only a minority of sera from hospitalized individuals lose more than threefold

© The Author(s), under exclusive license to Springer Nature Switzerland AG 2021   101
D. Focosi, *SARS-CoV-2 Spike Protein Convergent Evolution*,
SpringerBriefs in Microbiology,
https://doi.org/10.1007/978-3-030-87324-0_12

potency against any individual mutant, more than half of the mild/asymptomatic serum samples showed a threefold drop in potency against at least one Spike mutant [100]. While hyperimmune serum, monoclonal antibody and vaccine stockpiles could suddenly become ineffective and require months for the update, CCP collections can be immediately restarted and delivered an effective post-exposure prophylaxis and treatment in early disease stages.

References

1. Convertino I, Tuccori M, Ferraro S, Valdiserra G, Cappello E, Focosi D et al (2020) Exploring pharmacological approaches for managing cytokine storm associated with pneumonia and acute respiratory distress syndrome in COVID-19 patients. Crit Care (london, England) 24(1):331. https://doi.org/10.1186/s13054-020-03020-3
2. Tuccori M, Convertino I, Ferraro S, Cappello E, Valdiserra G, Focosi D et al (2020) The Impact of the COVID-19 "Infodemic" on drug-utilization behaviors: implications for pharmacovigilance. Drug Saf 43(8):699–709. https://doi.org/10.1007/s40264-020-00965-w
3. Group TRC (2020) Dexamethasone in hospitalized patients with covid-19—Preliminary report. N Engl J Med. https://doi.org/10.1056/NEJMoa2021436
4. Beigel JH, Tomashek KM, Dodd LE, Mehta AK, Zingman BS, Kalil AC et al (2020) Remdesivir for the treatment of covid-19—final report. N Engl J Med 383(19):1813–1826. https://doi.org/10.1056/NEJMoa2007764
5. Zhang S, Go EP, Ding H, Anang S, Kappes JC, Desaire H et al(2021a) Analysis of glycosylation and disulfide bonding of wild-type SARS-CoV-2 spike glycoprotein. https://doi.org/10.1101/2021.04.01.438120. %J bioRxiv
6. Harbison AM, Fogarty CA, Phung TK, Satheesan A, Schulz BL, Fadda E (2021) Fine-tuning the Spike: Role of the nature and topology of the glycan shield structure and dynamics of SARS-CoV-2 S. https://doi.org/10.1101/2021.04.01.438036. %J bioRxiv
7. MacGowan SA, Barton MI, Kutuzov M, Dushek O, van der Merwe PA, Barton GJ (2021) Missense variants in human ACE2 modify binding to SARS-CoV-2 Spike. https://doi.org/10.1101/2021.05.21.445118. %J bioRxiv
8. Yao H, Song Y, Chen Y, Wu N, Xu J, Sun C et al (2020) Molecular architecture of the SARS-CoV-2 virus. Cell 183(3):730–8.e13. https://doi.org/10.1016/j.cell.2020.09.018
9. Turoňová B, Sikora M, Schürmann C, Hagen WJH, Welsch S, Blanc FEC et al (2020) In situ structural analysis of SARS-CoV-2 spike reveals flexibility mediated by three hinges. Science 370(6513):203–208. https://doi.org/10.1126/science.abd5223
10. Kastenhuber ER, Jaimes JA, Johnson JL, Mercadante M, Muecksch F, Weisblum Y et al (2021) Coagulation factors directly cleave SARS-CoV-2 spike and enhance viral entry. https://doi.org/10.1101/2021.03.31.437960. %J bioRxiv
11. Wajnberg A, Amanat F, Firpo A, Altman DR, Bailey MJ, Mansour M et al (2020) Robust neutralizing antibodies to SARS-CoV-2 infection persist for months. Science 370(6521):1127–1130. https://doi.org/10.1126/science.abd7728
12. Yuan M, Huang D, Lee C-CD, Wu NC, Jackson AM, Zhu X et al (2021) Structural and functional ramifications of antigenic drift in recent SARS-CoV-2 variants. https://doi.org/10.1101/2021.02.16.430500. %J bioRxiv

© The Author(s), under exclusive license to Springer Nature Switzerland AG 2021
D. Focosi, *SARS-CoV-2 Spike Protein Convergent Evolution*,
SpringerBriefs in Microbiology,
https://doi.org/10.1007/978-3-030-87324-0

13. Klassen SA, Senefeld JW, Johnson PW, Carter RE, Wiggins CC, Shoham S et al (2020) Evidence favoring the efficacy of convalescent plasma for COVID-19 therapy. medRxiv : the preprint server for health sciences. https://doi.org/10.1101/2020.07.29.20162917

14. Klassen SA, Senefeld JW, Senese KA, Johnson PW, Wiggins CC, Baker SE et al (2021) Convalescent plasma therapy for COVID-19: a graphical mosaic of the worldwide evidence. Front Med8:684151. https://doi.org/10.3389/fmed.2021.684151

15. Focosi D, Farrugia A (2020) The art of the possible in approaching efficacy trialss for COVID19 convalescent plasma. Int J Infect Dis: IJID: off Publ Int Soc Infect Dis. https://doi.org/10.1016/j.ijid.2020.10.074

16. Tuccori M, Ferraro S, Convertino I, Cappello E, Valdiserra G, Blandizzi C et al (2020) Anti-SARS-CoV-2 neutralizing monoclonal antibodies: clinical pipeline. mAbs. 12(1):1854149. https://doi.org/10.1080/19420862.2020.1854149

17. Focosi D, Franchini M, Tuccori M (2020) The road towards polyclonal anti-SARS-CoV-2 immunoglobulins (hyperimmune serum) for passive immunization in COVID19. Am J Pathology

18. Li Q, Wu J, Nie J, Zhang L, Hao H, Liu S et al (2020) The Impact of mutations in SARS-CoV-2 spike on viral infectivity and antigenicity. Cell 182(5):1284–94.e9. https://doi.org/10.1016/j.cell.2020.07.012

19. Poland GA, Ovsyannikova IG, Kennedy RB (2020) SARS-CoV-2 immunity: review and applications to phase 3 vaccine candidates. Lancet. https://doi.org/10.1016/S0140-6736(20)32137-1

20. Liao H-Y, Huang H-Y, Chen X, Cheng C-W, Wang S-W, Shahed-Al-Mahmud M et al (2021) Impact of glycosylation on SARS-CoV-2 infection and broadly protective vaccine design. https://doi.org/10.1101/2021.05.25.445523. %J bioRxiv

21. Gorbalenya AE, Enjuanes L, Ziebuhr J, Snijder EJ (2006) Nidovirales: evolving the largest RNA virus genome. Virus Res 117(1):17–37. https://doi.org/10.1016/j.virusres.2006.01.017

22. Robson F, Khan KS, Le TK, Paris C, Demirbag S, Barfuss P et al (2020) Coronavirus RNA proofreading: molecular basis and therapeutic targeting. Mol Cell 79(5):710–727. https://doi.org/10.1016/j.molcel.2020.07.027

23. Gallaher WR (2020) A palindromic RNA sequence as a common breakpoint contributor to copy-choice recombination in SARS-COV-2. Adv Virol 165(10):2341–2348. https://doi.org/10.1007/s00705-020-04750-z

24. Bell SM, Katzelnick L, Bedford T (2019) Dengue genetic divergence generates within-serotype antigenic variation, but serotypes dominate evolutionary dynamics. eLife. 8. https://doi.org/10.7554/eLife.42496

25. Vilar S, Isom DG (2020) One year of SARS-CoV-2: how much has the virus changed? https://doi.org/10.1101/2020.12.16.423071. %J bioRxiv

26. Mercatelli D, Giorgi FM (2020) Geographic and genomic distribution of SARS-CoV-2 mutations. 11(1800). https://doi.org/10.3389/fmicb.2020.01800

27. Azgari C, Kilinc Z, Turhan B, Circi D, Adebali O (2021) The mutation profile of SARS-CoV-2 is primarily shaped by the host antiviral defense. https://doi.org/10.1101/2021.02.02.429486. %J bioRxiv

28. De Maio N, Walker CR, Turakhia Y, Lanfear R, Corbett-Detig R, Goldman N (2021) Mutation rates and selection on synonymous mutations in SARS-CoV-2. https://doi.org/10.1101/2021.01.14.426705. %J bioRxiv

29. Wu H, Xing N, Meng K, Fu B, Xue W, Dong P et al (2021a) Nucleocapsid mutation R203K/G204R increases the infectivity, fitness and virulence of SARS-CoV-2. https://doi.org/10.1101/2021.05.24.445386. %J bioRxiv

30. Niesen M, Anand P, Silvert E, Suratekar R, Pawlowski C, Ghosh P et al (2021) COVID-19 vaccines dampen genomic diversity of SARS-CoV-2: Unvaccinated patients exhibit more antigenic mutational variance. https://doi.org/10.1101/2021.07.01.21259833. %J medRxiv

31. Phelan J, Deelder W, Ward D, Campino S, Hibberd ML, Clark TG (2020) Controlling the SARS-CoV-2 outbreak, insights from large scale whole genome sequences generated across the world. https://doi.org/10.1101/2020.04.28.066977. %J bioRxiv

32. Tang X, Wu C, Li X, Song Y, Yao X, Wu X et al (2020) On the origin and continuing evolution of SARS-CoV-2. Natl Sci Rev 7(6):1012–1023. https://doi.org/10.1093/nsr/nwaa036%JNatio nalScienceReview

33. Alm E, Broberg EK, Connor T, Hodcroft EB, Komissarov AB, Maurer-Stroh S et al (2020) Geographical and temporal distribution of SARS-CoV-2 clades in the WHO European Region, January to June 2020. https://doi.org/10.2807/1560-7917.ES.2020.25.32.2001410

34. Rambaut A, Holmes EC, O'Toole Á, Hill V, McCrone JT, Ruis C et al (2020a) A dynamic nomenclature proposal for SARS-CoV-2 lineages to assist genomic epidemiology. Nat Microbiol 5(11):1403–1407. https://doi.org/10.1038/s41564-020-0770-5

35. Pybus O: Pango (2021) Lineage Nomenclature: provisional rules for naming recombinant lineages. https://virological.org/t/pango-lineage-nomenclature-provisional-rules-for-naming-recombinant-lineages/657. Accessed 17 March 2021

36. Hadfield J, Megill C, Bell SM, Huddleston J, Potter B, Callender C et al (2018) Nextstrain: real-time tracking of pathogen evolution. Bioinformatics 34(23):4121–4123. https://doi.org/10.1093/bioinformatics/bty407%JBioinformatics

37. Bedford T, Hodcroft EN (2021) RA: updated Nextstain SARS-CoV-2 clade naming strategy. https://virological.org/t/updated-nextstain-sars-cov-2-clade-naming-strategy/581

38. Jacob JJ, Vasudevan K, Pragasam AK, Gunasekaran K, Kang G, Veeraraghavan B et al (2020) Evolutionary tracking of SARS-CoV-2 genetic variants highlights intricate balance of stabilizing and destabilizing mutations. https://doi.org/10.1101/2020.12.22.423920. %J bioRxiv

39. Young BE, Fong SW, Chan YH, Mak TM, Ang LW, Anderson DE et al (2020) Effects of a major deletion in the SARS-CoV-2 genome on the severity of infection and the inflammatory response: an observational cohort study. Lancet 396(10251):603–611. https://doi.org/10.1016/s0140-6736(20)31757-8

40. Koçhan N, Eskier D, Suner A, Karakülah G, Oktay Y (2021) Different selection dynamics of S and RdRp between SARS-CoV-2 genomes with and without the dominant mutations. https://doi.org/10.1101/2021.01.03.20237602 %J medRxiv

41. Cotten M, Robertson DL, Phan MVT (2021) Unique protein features of SARS-CoV-2 relative to other Sarbecoviruses. https://doi.org/10.1101/2021.04.06.438675. %J bioRxiv

42. Ruan Y, Hou M, Li J, Song Y, Wang H-Y, Zeng H et al (2021) One viral sequence for each host?—The neglected within-host diversity as the main stage of SARS-CoV-2 evolution. https://doi.org/10.1101/2021.06.21.449205. %J bioRxiv

43. Schroers B, Gudimella R, Bukur T, Roesler T, Loewer M, Sahin U (2021) Large-scale analysis of SARS-CoV-2 spike-glycoprotein mutants demonstrates the need for continuous screening of virus isolates. https://doi.org/10.1101/2021.02.04.429765. %J bioRxiv

44. Tonkin-Hill G, Martincorena I, Amato R, Lawson ARJ, Gerstung M, Johnston I et al (2020) Patterns of within-host genetic diversity in SARS-CoV-https://doi.org/10.1101/2020.12.23.424229. %J bioRxiv

45. Takada K, Ueda MT, Watanabe T, Nakagawa S (2020) Genomic diversity of SARS-CoV-2 can be accelerated by a mutation in the nsp14 gene. https://doi.org/10.1101/2020.12.23.424231. %J bioRxiv

46. Rao RSP, Ahsan N, Xu C, Su L, Verburgt J, Fornelli L et al (2021) Evolutionary dynamics of indels in SARS-CoV-2 spike glycoprotein. https://doi.org/10.1101/2021.07.30.454557. %J bioRxiv

47. McCarthy KR, Rennick LJ, Nambulli S, Robinson-McCarthy LR, Bain WG, Haidar G et al (2020) Natural deletions in the SARS-CoV-2 spike glycoprotein drive antibody escape. https://doi.org/10.1101/2020.11.19.389916. %J bioRxiv

48. Kemp S, Datir R, Collier D, Ferreira I, Carabelii A, Harvey W et al (2020a) Recurrent emergence and transmission of a SARS-CoV-2 Spike deletion ΔH69/V70. https://doi.org/10.1101/2020.12.14.422555. %J bioRxiv

49. Resende P, Naveca F, Lins R, Zimmer Dezordi F, Ferraz M, Moreira E et al (2021a) The ongoing evolution of variants of concern and interest of SARS-CoV-2 in Brazil revealed by convergent indels in the amino (N)-terminal domain of the Spike protein. https://vir

ological.org/t/the-ongoing-evolution-of-variants-of-concern-and-interest-of-sars-cov-2-in-brazil-revealed-by-convergent-indels-in-the-amino-n-terminal-domain-of-the-spike-protei n/659. Accessed 20 Mar 2021

50. Gupta K, Toelzer C, Kavanagh Williamson M, Shoemark DK, Oliveira ASF, Matthews DA et al (2021a) Structure and mechanism of SARS-CoV-2 Spike N679-V687 deletion variant elucidate cell-type specific evolution of viral fitness. https://doi.org/10.1101/2021.05.11.443384. %J bioRxiv

51. Garushyants SK, Rogozin IB, Koonin EV (2021) Insertions in SARS-CoV-2 genome caused by template switch and duplications give rise to new variants of potential concern. https://doi.org/10.1101/2021.04.23.441209. %J bioRxiv

52. Varabyou A, Pockrandt C, Salzberg SL, Pertea M (2020) Rapid detection of inter-clade recombination in SARS-CoV-2 with Bolotie. https://doi.org/10.1101/2020.09.21.300913. %J bioRxiv

53. VanInsberghe D, Neish A, Lowen AC, Koelle K (2020) Identification of SARS-CoV-2 recombinant genomes. https://doi.org/10.1101/2020.08.05.238386. %J bioRxiv

54. De Maio N, Walker C, Borges R, Weilguny L (2020) Issues with SARS-CoV-2 sequencing data. https://virological.org/t/issues-with-sars-cov-2-sequencing-data/473

55. van Dorp L, Richard D, Tan CCS, Shaw LP, Acman M, Balloux F (2020a) No evidence for increased transmissibility from recurrent mutations in SARS-CoV-2. Nat Commun 11(1):5986. https://doi.org/10.1038/s41467-020-19818-2

56. Nie Q, Li X, Chen W, Liu D, Chen Y, Li H et al (2020) Phylogenetic and phylodynamic analyses of SARS-CoV-2. Virus Res. https://doi.org/10.1016/j.virusres.2020.198098

57. Wang H, Kosakovsky Pond S, Nekrutenko A, Nielsen R (2020) Testing recombination in the pandemic SARS-CoV-2 strains. https://virological.org/t/testing-recombination-inthe-pandemic-sars-cov-2-strains/492

58. Richard D, Owen CJ, van Dorp L, Balloux F (2020a) No detectable signal for ongoing genetic recombination in SARS-CoV-2. https://doi.org/10.1101/2020.12.15.422866. %J bioRxiv

59. Ignatieva A, Hein J, Jenkins PA (2021) Evidence of ongoing recombination in SARS-CoV-2 through genealogical reconstruction. https://doi.org/10.1101/2021.01.21.427579. %J bioRxiv

60. Müller NF, Kistler K, Bedford T (2021a) Recombination patterns in coronaviruses. https://doi.org/10.1101/2021.04.28.441806. %J bioRxiv

61. Jackson B, Boni MF, Bull MJ, Colleran A, Colquhoun RM, Darby A et al (2021a) Generation and transmission of inter-lineage recombinants in the SARS-CoV-2 pandemic. https://doi.org/10.1101/2021.06.18.21258689. %J medRxiv

62. Chen Y-S, Lu S, Zhang B, Du T, Li W-J, Lei M et al (2021a) Comprehensive analysis of RNA-seq and whole genome sequencing data reveals no evidence for SARS-CoV-2 integrating into host genome. https://doi.org/10.1101/2021.06.06.447293. %J bioRxiv

63. Faria N, Claro I, Candido D, Franco L, Andrade P, Coletti T et al (2021) Genomic characterisation of an emergent SARS-CoV-2 lineage in Manaus: preliminary findings. https://virological.org/t/genomic-characterisation-of-an-emergent-sars-cov-2-lineage-in-manaus-preliminary-findings/586. Accessed 27 Jan 2021

64. Martin D, Weaver S, Tegally H, James San E, Wilkinson E, Giandhari J et al (2021) The emergence and ongoing convergent evolution of the N501Y lineages coincided with a major global shift in the SARS-Cov-2 selective landscape. https://virological.org/t/the-emergence-and-ongoing-convergent-evolution-of-the-n501y-lineages-coincided-with-a-major-global-shift-in-the-sars-cov-2-selective-landscape/618

65. Laskar R, Ali S (2021) Differential mutational profile of SARS-CoV-2 proteins across deceased and asymptomatic patients. https://doi.org/10.1101/2021.03.31.437815. %J bioRxiv

66. Pang X, Li P, Zhang L, Que L, Dong M, Wang Q et al (2021) Emerging SARS-CoV-2 mutation hotspots associated with clinical outcomes. https://doi.org/10.1101/2021.03.31.437666. %J bioRxiv

67. Nielsen BF, Eilersen A, Simonsen L, Sneppen K (2021) Lockdowns exert selection pressure on overdispersion of SARS-CoV-2 variants. https://doi.org/10.1101/2021.06.30.21259771. %J medRxiv

68. Pereson MJ, Mojsiejczuk L, Martínez AP, Flichman DM, Garcia GH, Di Lello FA. Phylogenetic analysis of SARS-CoV-2 in the first few months since its emergence.n/a(n/a). https://doi.org/10.1002/jmv.26545.

69. van Dorp L, Acman M, Richard D, Shaw LP, Ford CE, Ormond L et al (2020b) Emergence of genomic diversity and recurrent mutations in SARS-CoV-2. Infect Genet Evol 83:104351. https://doi.org/10.1016/j.meegid.2020.104351

70. Liu Q, Zhao S, Shi C-M, Song S, Zhu S, Su Y et al (2020a) Population genetics of SARS-CoV-2: disentangling effects of sampling bias and infection clusters. Genom Proteomics Bioinform. https://doi.org/10.1016/j.gpb.2020.06.001

71. Pereson MJ, Flichman DM, Martínez AP, Baré P, Garcia GH, DI Lello FA (2020) Evolutionary analysis of SARS-CoV-2 spike protein for its different clades. https://doi.org/10.1101/2020.11.24.396671. %J bioRxiv

72. Borges V, Alves MJ, Amicone M, Isidro J, Ze-Ze L, Duarte S et al (2021a) Mutation rate of SARS-CoV-2 and emergence of mutators during experimental evolution. https://doi.org/10.1101/2021.05.19.444774. %J bioRxiv

73. Jolly B, Rophina M, Shamnath A, Imran M, Bhoyar RC, Divakar MK et al (2020) Genetic epidemiology of variants associated with immune escape from global SARS-CoV-2 genomes. https://doi.org/10.1101/2020.12.24.424332. %J bioRxiv

74. Garry R, Andersen K, Gallaher W, Tsan-Yuk Lam T, Gangaparapu K, Latif A et al (2021) Spike protein mutations in novel SARS-CoV-2 'variants of concern' commonly occur in or near indels. https://virological.org/t/spike-protein-mutations-in-novel-sars-cov-2-variants-of-concern-commonly-occur-in-or-near-indels/605

75. Hoffmann D, Mereiter S, Oh YJ, Monteil V, Zhu R, Canena D et al (2021a) Identification of lectin receptors for conserved SARS-CoV-2 glycosylation sites. https://doi.org/10.1101/2021.04.01.438087. %J bioRxiv

76. Correa Giron C, Laaksonen A, Barroso da Silva FL (2021) The Up state of the SARS-COV-2 Spike homotrimer favors an increased virulence for new variants. https://doi.org/10.1101/2021.04.05.438465. %J bioRxiv

77. Korber B, Fischer WM, Gnanakaran S, Yoon H, Theiler J, Abfalterer W et al (2020) Tracking changes in SARS-CoV-2 spike: evidence that D614G increases infectivity of the COVID-19 virus. Cell 182(4):812–27.e19. https://doi.org/10.1016/j.cell.2020.06.043

78. Trucchi E, Gratton P, Mafessoni F, Motta S, Cicconardi F, Bertorelle G et al (2020) Unveiling diffusion pattern and structural impact of the most invasive SARS-CoV-2 spike mutation. https://doi.org/10.1101/2020.05.14.095620. %J bioRxiv

79. Guo C, Tsai SJ, Ai Y, Li M, Pekosz A, Cox A et al (2020) The D614G Mutation Enhances the lysosomal trafficking of SARS-CoV-2 spike. https://doi.org/10.1101/2020.12.08.417022. %J bioRxiv

80. Plante JA, Liu Y, Liu J, Xia H, Johnson BA, Lokugamage KG et al (2020) Spike mutation D614G alters SARS-CoV-2 fitness. Nature. https://doi.org/10.1038/s41586-020-2895-3

81. Zhang L, Jackson CB, Mou H, Ojha A, Rangarajan ES, Izard T et al (2020a) The D614G mutation in the SARS-CoV-2 spike protein reduces S1 shedding and increases infectivity. https://doi.org/10.1101/2020.06.12.148726. %J bioRxiv

82. Cheng Y-W, Chao T-L, Li C-L, Wang S-H, Kao H-C, Tsai Y-M et al (2021a) D614G substitution of SARS-CoV-2 spike protein increases syncytium formation and viral transmission via enhanced furin-mediated spike cleavage. https://doi.org/10.1101/2021.01.27.428541. %J bioRxiv

83. Bamberger TC, Pankow S, Martinez-Bartolome S, Diedrich J, Park R, Yates JR. The host interactome of spike expands the tropism of SARS-CoV-2. https://doi.org/10.1101/2021.02.16.431318. %J bioRxiv

84. Yang T-J, Yu P-Y, Chang Y-C, Kuo C-W, Khoo K-H, Hsu S-TD (2021a) COVID-19 dominant D614G mutation in the SARS-CoV-2 spike protein desensitizes its temperature-dependent denaturation. https://doi.org/10.1101/2021.03.28.437426. %J bioRxiv

85. Lee CY-P, Amrun SN, Chee RS-L, Goh YS, Mak TM, Octavia S et al (2020) Neutralizing antibodies from early cases of SARS-CoV-2 infection offer cross-protection against the SARS-CoV-2 D614G variant. https://doi.org/10.1101/2020.10.08.332544. %J bioRxiv

86. Weissman D, Alameh M-G, de Silva T, Collini P, Hornsby H, Brown R et al (2020) D614G Spike mutation increases SARS CoV-2 susceptibility to neutralization. https://doi.org/10.1101/2020.07.22.20159905 %J medRxiv.

87. Kepler L, Hamins-Puertolas M, Rasmussen DA (2020) Decomposing the sources of SARS-CoV-2 fitness variation in the United States. https://doi.org/10.1101/2020.12.14.422739. %J bioRxiv

88. Zhang BZ, Hu YF, Chen LL, Yau T, Tong YG, Hu JC et al (2020b) Mining of epitopes on spike protein of SARS-CoV-2 from COVID-19 patients. Cell Res 30(8):702–704. https://doi.org/10.1038/s41422-020-0366-x

89. Hodcroft EB, Zuber M, Nadeau S, Comas I, González Candelas F, Stadler T et al (2020) Emergence and spread of a SARS-CoV-2 variant through Europe in the summer of 2020. https://doi.org/10.1101/2020.10.25.20219063 %J medRxiv.

90. Grabowski F, Kochanczyk M, Lipniacki T (2021a) L18F substrain of SARS-CoV-2 VOC-202012/01 is rapidly spreading in England. https://doi.org/10.1101/2021.02.07.21251262 %J medRxiv.

91. Severinov D, Yi B, Dalpke AR, F, Winkler S, Beil JR, S, Klemroth SM, Hölzer G, Kühnert M, Poetschm AR, et al (2021) Potential novel SARS-CoV-2 variant of interest from lineage AP.1 with N501Y and additional spike mutations identified in Saxony, Germany. https://virological.org/t/potential-novel-sars-cov-2-variant-of-interest-from-lineage-ap-1-with-n501y-and-additional-spike-mutations-identified-in-saxony-germany/674

92. Bal A, Destras G, Gaymard A, Regue H, Semanas Q, d'Aubarde C et al (2020) Two-step strategy for the identification of SARS-CoV-2 variants co-occurring with spike deletion H69-V70, Lyon, France, August to December 2020. https://doi.org/10.1101/2020.11.10.20228528 %J medRxivs

93. Washington NL, White S, Schiabor Barrett KM, Cirulli ET, Bolze A, Lu JT (2020) S gene dropout patterns in SARS-CoV-2 tests suggest spread of the H69del/V70del mutation in the US. https://doi.org/10.1101/2020.12.24.20248814 %J medRxiv.

94. Gravagnuolo AM, Faqih L, Cronshaw C, Wynn J, Burglin L, Klapper P et al (2021) Epidemiological Investigation of New SARS-CoV-2 Variant of Concern 202012/01 in England. https://doi.org/10.1101/2021.01.14.21249386 %J medRxiv.

95. Sherrill-Mix S, Van Duyne GD, Bushman FD (2021) Molecular beacons allow specific RT-LAMP detection of B.1.1.7 variant SARS-CoV-2. https://doi.org/10.1101/2021.03.25.21254356 %J medRxiv.

96. Kidd M, Richter A, Best A, Mirza J, Percival B, Mayhew M et al (2020) S-variant SARS-CoV-2 is associated with significantly higher viral loads in samples tested by ThermoFisher TaqPath RT-PCR. https://doi.org/10.1101/2020.12.24.20248834 %J medRxiv.

97. Larsen B, Worobey M (2021) Identification of a novel SARS-CoV-2 Spike 69–70 deletion lineage circulating in the United States. https://virological.org/t/identification-of-a-novel-sars-cov-2-spike-69-70-deletion-lineage-circulating-in-the-united-states/577

98. Moreno G, Braun K, Larsen B, Alpert T, Worobey M, Grubaugh N et al (2021) Detection of non-B.1.1.7 Spike Δ69/70 sequences (B.1.375) in the United States. https://virological.org/t/detection-of-non-b-1-1-7-spike-69-70-sequences-b-1-375-in-the-united-states/587. Accessed 8 Feb 2021

99. Brejová B, Hodorová V, Boršová K, Čabanová V, Reizigová L, Paul E et al (2021) B.1.258Δ, a SARS-CoV-2 variant with ΔH69/ΔV70 in the Spike protein circulating in the Czech Republic and Slovakia. https://virological.org/t/b-1-258-a-sars-cov-2-variant-with-h69-v70-in-the-spike-protein-circulating-in-the-czech-republic-and-slovakia/613. Accessed 8 Feb 2021

100. Rees-Spear C, Muir L, Griffith SA, Heaney J, Aldon Y, Snitselaar J et al (2021) The impact of Spike mutations on SARS-CoV-2 neutralization. https://doi.org/10.1101/2021.01.15.426849. %J bioRxiv. Accessed 5 Feb 2021

101. Cherian S, Potdar V, Jadhav S, Yadav P, Gupta N, Das M et al (2021) Convergent evolution of SARS-CoV-2 spike mutations, L452R, E484Q and P681R, in the second wave of COVID-19 in Maharashtra, India. https://doi.org/10.1101/2021.04.22.440932. %J bioRxiv

102. Kubik S, Arrigo N, Bonet J, Xu Z (2021) Mutational hotspot in the SARS-CoV-2 Spike protein N-terminal domain conferring immune escape potential. https://doi.org/10.1101/2021.05.28. 446137. %J bioRxiv

103. Yang T-J, Yu P-Y, Chang Y-C, Liang K-H, Tso H-C, Ho M-R et al (2021b) Impacts on the structure-function relationship of SARS-CoV-2 spike by B.1.1.7 mutations. https://doi.org/10.1101/2021.05.11.443686. %J bioRxiv

104. Deng X, Garcia-Knight MA, Khalid MM, Servellita V, Wang C, Morris MK et al (2021) Transmission, infectivity, and neutralization of a Spike L452R SARS-CoV-2 variant. Cell. https://doi.org/10.1016/j.cell.2021.04.025

105. Tegally H, Wilkinson E, Giovanetti M, Iranzadeh A, Fonseca V, Giandhari J et al (2020) Emergence and rapid spread of a new severe acute respiratory syndrome-related coronavirus 2 (SARS-CoV-2) lineage with multiple spike mutations in South Africa. https://doi.org/10.1101/2020.12.21.20248640 %J medRxiv.

106. Hoffmann M, Kleine-Weber H, Pöhlmann S (2020) A Multibasic cleavage site in the spike protein of SARS-CoV-2 is essential for infection of human lung cells. Mol Cell 78(4):779–84.e5. https://doi.org/10.1016/j.molcel.2020.04.022

107. Peacock TP, Goldhill DH, Zhou J, Baillon L, Frise R, Swann OC et al (2020) The furin cleavage site of SARS-CoV-2 spike protein is a key determinant for transmission due to enhanced replication in airway cells. https://doi.org/10.1101/2020.09.30.318311. %J bioRxiv

108. Zhu Y, Feng F, Hu G, Wang Y, Yu Y, Zhu Y et al (2020) The S1/S2 boundary of SARS-CoV-2 spike protein modulates cell entry pathways and transmission. https://doi.org/10.1101/2020.08.25.266775. %J bioRxiv

109. Happi C, Ihekweazu C, Nkengasong J, Eniola Oluniyi P, Olawoye I (2020) Detection of SARS-CoV-2 P681H Spike Protein Variant in Nigeria. https://virological.org/t/detection-of-sars-cov-2-p681h-spike-protein-variant-in-nigeria/567

110. Zuckerman NS, Fleishon S, Bucris E, Bar-Ilan D, Linial M, Bar-Or I et al (2021) A unique SARS-CoV-2 spike protein P681H strain detected in Israel. https://doi.org/10.1101/2021.03.25.21253908 %J medRxiv.

111. Lasek-Nesselquist E, Pata J, Schneider E, St. George K (2021) A tale of three SARS-CoV-2 variants with independently acquired P681H mutations in New York State. https://doi.org/10.1101/2021.03.10.21253285 %J medRxiv.

112. Lubinski B, Tang T, Daniel S, Jaimes JA, Whittaker G (2021) Functional evaluation of proteolytic activation for the SARS-CoV-2 variant B.1.1.7: role of the P681H mutation. https://doi.org/10.1101/2021.04.06.438731. %J bioRxiv

113. Peacock TP, Sheppard CM, Brown JC, Goonawardane N, Zhou J, Whiteley M et al (2021) The SARS-CoV-2 variants associated with infections in India, B.1.617, show enhanced spike cleavage by furin. https://doi.org/10.1101/2021.05.28.446163. %J bioRxiv

114. Saito A, Nasser H, Uriu K, Kosugi Y, Irie T, Shirakawa K et al (2021) SARS-CoV-2 spike P681R mutation enhances and accelerates viral fusion. https://doi.org/10.1101/2021.06.17.448820. %J bioRxiv

115. Frazier LE, Lubinski B, Tang T, Daniel S, Jaimes JA, Whittaker GR (2021) Spike protein cleavage-activation mediated by the SARS-CoV-2 P681R mutation: a case-study from its first appearance in variant of interest (VOI) A.23.1 identified in Uganda. https://doi.org/10.1101/2021.06.30.450632. %J bioRxiv

116. Molina-Mora JA (2021) Insights into the mutation T1117I in the spike and the lineage B.1.1.389 of SARS-CoV-2 circulating in Costa Rica. https://doi.org/10.1101/2021.07.08.451640. %J bioRxiv

117. Brouwer PJM, Caniels TG, van der Straten K, Snitselaar JL, Aldon Y, Bangaru S et al (2020) Potent neutralizing antibodies from COVID-19 patients define multiple targets of vulnerability. Science 369(6504):643–650. https://doi.org/10.1126/science.abc5902

118. Huang K-YA, Tan TK, Chen T-H, Huang C-G, Harvey R, Hussain S et al (2020) Plasmablast-derived antibody response to acute SARS-CoV-2 infection in humans. https://doi.org/10.1101/2020.08.28.267526. %J bioRxiv

119. Seydoux E, Homad LJ, MacCamy AJ, Parks KR, Hurlburt NK, Jennewein MF et al (2020) Analysis of a SARS-CoV-2-infected individual reveals development of potent neutralizing antibodies with limited somatic mutation. Immunity 53(1):98-105.e5. https://doi.org/10.1016/j.immuni.2020.06.001

120. Voss WN, Hou YJ, Johnson NV, Kim JE, Delidakis G, Horton AP et al (2020) Prevalent, protective, and convergent IgG recognition of SARS-CoV-2 non-RBD spike epitopes in COVID-19 convalescent plasma. https://doi.org/10.1101/2020.12.20.423708. %J bioRxiv

121. Piccoli L, Park YJ, Tortorici MA, Czudnochowski N, Walls AC, Beltramello M et al (2020) Mapping neutralizing and immunodominant sites on the SARS-CoV-2 spike receptor-binding domain by structure-guided high-resolution serology. Cell 183(4):1024–42.e21. https://doi.org/10.1016/j.cell.2020.09.037

122. Steffen TL, Stone ET, Hassert M, Geerling E, Grimberg BT, Espino AM et al (2020) The receptor binding domain of SARS-CoV-2 spike is the key target of neutralizing antibody in human polyclonal sera. https://doi.org/10.1101/2020.08.21.261727. %J bioRxiv

123. Greaney AJ, Loes AN, Crawford KH, Starr TN, Malone KD, Chu HY et al (2021a) Comprehensive mapping of mutations to the SARS-CoV-2 receptor-binding domain that affect recognition by polyclonal human serum antibodies. https://www.biorxiv.org/content/biorxiv/early/2021/01/04/2020.12.31.425021.full.pdf. Accessed 27 2021

124. Ortega JT, Serrano ML, Pujol FH, Rangel HR (2020) Role of changes in SARS-CoV-2 spike protein in the interaction with the human ACE2 receptor: an in silico analysis. EXCLI J 19:410–417. https://doi.org/10.17179/excli2020-1167

125. Rockx B, Donaldson E, Frieman M, Sheahan T, Corti D, Lanzavecchia A et al (2010) Escape from human monoclonal antibody neutralization affects in vitro and in vivo fitness of severe acute respiratory syndrome coronavirus. J Infect Dis 201(6):946–955. https://doi.org/10.1086/651022

126. Teng S, Sobitan A, Rhoades R, Liu D, Tang Q (2020) Systemic effects of missense mutations on SARS-CoV-2 spike glycoprotein stability and receptor-binding affinity. Brief Bioinform. https://doi.org/10.1093/bib/bbaa233

127. Chakraborty S (2021) Evolutionary and structural analysis elucidates mutations on SARS-CoV2 spike protein with altered human ACE2 binding affinity. Biochem Biophys Res Commun 534:374–380. https://doi.org/10.1016/j.bbrc.2020.11.075

128. Lule Bugembe D, Phan MVT, Ssewanyana I, Semanda P, Nansumba H, Dhaala B et al (2020) A SARS-CoV-2 lineage A variant (A.23.1) with altered spike has emerged and is dominating the current Uganda epidemic. https://doi.org/10.1101/2021.02.08.21251393 %J medRxiv

129. Zech F, Schniertshauer D, Jung C, Herrmann A, Xie Q, Nchioua R et al (2021) Spike mutation T403R allows bat coronavirus RaTG13 to use human ACE2. https://doi.org/10.1101/2021.05.31.446386. %J bioRxiv

130. Villoutreix B, Calvez V, Marcelin A-G, Khatib A-M (2021) In silico investigation of the new UK (B.1.1.7) and South African (501Y.V2) SARS-CoV-2 variants with a focus at the ACE2-Spike RBD interface. https://doi.org/10.1101/2021.01.24.427939. %J bioRxiv

131. Wang R, Zhang Q, Ge J, Ren W, Zhang R, Lan J et al (2021a) Spike mutations in SARS-CoV-2 variants confer resistance to antibody neutralization. https://doi.org/10.1101/2021.03.09.434497. %J bioRxiv

132. Laffeber C, de Koning K, Kanaar R, Lebbink J (2021) Experimental evidence for enhanced receptor binding by rapidly spreading SARS-CoV-2 variants. https://doi.org/10.1101/2021.02.22.432357. %J bioRxiv

133. Starr TN, Greaney AJ, Dingens AS, Bloom JD (2021a) Complete map of SARS-CoV-2 RBD mutations that escape the monoclonal antibody LY-CoV555 and its cocktail with LY-CoV016. Cel Rep Med 2(4):100255. https://doi.org/10.1101/2021.02.17.431683. %J bioRxiv

134. Luan B, Huynh T (2021) Insights on SARS-CoV-2's mutations for evading human antibodies: sacrifice and survival. https://doi.org/10.1101/2021.02.06.430088. %J bioRxiv

135. UK (2017) Guidelines from the expert advisory committee on the Safety of Blood, Tissues and Organs (SaBTO) on measures to protect patients from acquiring hepatitis E virus via transfusion or transplantation

136. Starr TN, Greaney AJ, Addetia A, Hannon WW, Choudhary MC, Dingens AS et al (2021b) Prospective mapping of viral mutations that escape antibodies used to treat COVID-19. Science 2021:eabf9302. https://doi.org/10.1101/2020.11.30.405472. %J bioRxiv

137. Greaney AJ, Starr TN, Gilchuk P, Zost SJ, Binshtein E, Loes AN et al (2020) Complete mapping of mutations to the SARS-CoV-2 spike receptor-binding domain that escape antibody recognition. Cell Host Microbe. https://doi.org/10.1016/j.chom.2020.11.007

138. Copin R, Baum A, Wloga E, Pascal KE, Giordano S, Fulton BO et al (2021) REGEN-COV protects against viral escape in preclinical and human studies. https://doi.org/10.1101/2021.03.10.434834. %J bioRxiv

139. Wang Z, Schmidt F, Weisblum Y, Muecksch F, Barnes CO, Finkin S et al (2021) mRNA vaccine-elicited antibodies to SARS-CoV-2 and circulating variants. https://www.biorxiv.org/content/biorxiv/early/2021/01/19/2021.01.15.426911.full.pdf

140. Yi C, Sun X, Ye J, Ding L, Liu M, Yang Z et al (2020) Key residues of the receptor binding motif in the spike protein of SARS-CoV-2 that interact with ACE2 and neutralizing antibodies. Cell Mol Immunol 17(6):621–630. https://doi.org/10.1038/s41423-020-0458-z

141. Thomson EC, Rosen LE, Shepherd JG, Spreafico R, da Silva Filipe A, Wojcechowskyj JA et al (2020) The circulating SARS-CoV-2 spike variant N439K maintains fitness while evading antibody-mediated immunity. https://doi.org/10.1101/2020.11.04.355842. %J bioRxiv

142. Starr TN, Greaney AJ, Hilton SK, Ellis D, Crawford KHD, Dingens AS et al (2020) Deep mutational scanning of SARS-CoV-2 receptor binding domain reveals constraints on folding and ACE2 Binding. Cell 182(5):1295–310.e20. https://doi.org/10.1016/j.cell.2020.08.012

143. Weisblum Y, Schmidt F, Zhang F, DaSilva J, Poston D, Lorenzi JCC et al (2020) Escape from neutralizing antibodies by SARS-CoV-2 spike protein variants. eLife 28(9):e61312. https://doi.org/10.7554/eLife.61312.

144. Barnes CO, Jette CA, Abernathy ME, Dam K-MA, Esswein SR, Gristick HB et al (2020) SARS-CoV-2 neutralizing antibody structures inform therapeutic strategies. Nature 588(7839):682–687. https://doi.org/10.1038/s41586-020-2852-1

145. Schrörs B, Gudimella R, Bukur T, Rösler T, Löwer M, Sahin U (2021) Large-scale analysis of SARS-CoV-2 spike-glycoprotein mutants demonstrates the need for continuous screening of virus isolates. https://doi.org/10.1101/2021.02.04.429765. %J bioRxiv

146. Garcia V, Vig V, Peillard L, Ramdani A, Mohamed S, Halfon P (2021) First description of two immune escape indian B.1.1.420 and B.1.617.1 SARS-CoV2 variants in France. https://doi.org/10.1101/2021.05.12.443357. %J bioRxiv

147. Pattabiraman C, Prasad P, George AK, Sreenivas D, Rasheed R, Reddy NVK et al (2021) Importation, circulation, and emergence of variants of SARS-CoV-2 in the South Indian State of Karnataka. https://doi.org/10.1101/2021.03.17.21253810 %J medRxiv.

148. Singh J, Ehtesham NZ, Rahman SA, Hasnain SE (2021) Structure-function investigation of a new VUI-202012/01 SARS-CoV-2 variant. https://doi.org/10.1101/2021.01.01.425028. %J bioRxiv

149. Gupta V, Bhoyar RC, Jain A, Srivastava S, Upadhayay R, Imran M et al (2020) Asymptomatic reinfection in two healthcare workers from India with genetically distinct SARS-CoV-2. Clinical infectious diseases : an official publication of the Infectious Diseases Society of America. https://doi.org/10.1093/cid/ciaa1451

150. Rani P, Imran M, Lakshmi J, Jolly B, Jain AS, A, Senthivel V et al (2021) Symptomatic reinfection of SARS-CoV-2 with spike protein variant N440K associated with immune escape. https://osf.io/7gk69/. Accessed 14 Mar 2021

151. Islam SR, Prusty D, Manna SK (2021) Structural basis of fitness of emerging SARS-COV-2 variants and considerations for screening, testing and surveillance strategy to contain their threat. https://doi.org/10.1101/2021.01.28.21250666 %J medRxiv.

152. Zrelovs N, Ustinova M, Silamiķelis I, Birzniece L, Megnis K, Rovīte V et al(2020) First report on the Latvian SARS-CoV-2 isolate genetic diversity. https://doi.org/10.1101/2020.09.08.20190504 %J medRxiv.

153. Amanat F, Thapa M, Lei T, Ahmed SMS, Adelsberg DC, Carreno JM et al (2021) The plasmablast response to SARS-CoV-2 mRNA vaccination is dominated by non-neutralizing

antibodies that target both the NTD and the RBD. https://doi.org/10.1101/2021.03.07.212 53098 %J medRxiv.

154. Tandel D, Gupta D, Sah V, Harshan KH (2021) N440K variant of SARS-CoV-2 has Higher Infectious Fitness. https://doi.org/10.1101/2021.04.30.441434. %J bioRxiv

155. Gong SY, Chatterjee D, Richard J, Prevost J, Tauzin A, Gasser R et al (2021) Contribution of single mutations to selected SARS-CoV-2 emerging variants Spike antigenicity. https://doi.org/10.1101/2021.08.04.455140. %J bioRxiv

156. Tchesnokova V, Kulakesara H, Larson L, Bowers V, Rechkina E, Kisiela D et al (2021) Acquisition of the L452R mutation in the ACE2-binding interface of Spike protein triggers recent massive expansion of SARS-Cov-2 variants. https://doi.org/10.1101/2021.02.22.432189. %J bioRxiv

157. Captive gorillas test positive for coronavirus (2021). https://www.sciencemag.org/news/2021/01/captive-gorillas-test-positive-coronavirus

158. Zhang W, Davis B, Chen SS, Sincuir Martinez J, Plummer JT, Vail E (2021b) Emergence of a novel SARS-CoV-2 strain in Southern California, USA. JAMA. https://doi.org/10.1001/jama.2021.1612

159. Motozono C, Toyoda M, Zahradnik J, Ikeda T, Saito A, Tan TS et alAn emerging SARS-CoV-2 mutant evading cellular immunity and increasing viral infectivity. 2021:2021.04.02.438288. https://doi.org/10.1101/2021.04.02.438288. %J bioRxiv

160. Tada T, Zhou H, Dcosta BM, Samanovic MI, Mulligan MJ, Landau NR (2021a) SARS-CoV-2 Lambda Variant Remains Susceptible to Neutralization by mRNA Vaccine-elicited Antibodies and Convalescent Serum. https://doi.org/10.1101/2021.07.02.450959. %J bioRxiv

161. Alenquer M, Ferreira F, Lousa D, Valerio M, Medina-Lopes M, Bergman M-L et al (2021) Amino acids 484 and 494 of SARS-CoV-2 spike are hotspots of immune evasion affecting antibody but not ACE2 binding. https://doi.org/10.1101/2021.04.22.441007. %J bioRxiv

162. Koenig P-A, Das H, Liu H, Kümmerer BM, Gohr FN, Jenster L-M et al (2021) Structure-guided multivalent nanobodies block SARS-CoV-2 infection and suppress mutational escape. https://doi.org/10.1126/science.abe6230 %J Science

163. Fratev F (2020) The SARS-CoV-2 S1 spike protein mutation N501Y alters the protein interactions with both hACE2 and human derived antibody: A Free energy of perturbation study. https://doi.org/10.1101/2020.12.23.424283. %J bioRxiv

164. Zhang G, Liu H, Zhang Q, Wei P, Chen Z, Aviszus K et al (2021c) The basis of a more contagious 501Y.V1 variant of SARS-COV-2. https://doi.org/10.1101/2021.02.02.428884. %J bioRxiv

165. Tian F, Tong B, Sun L, Shi S, Zheng B, Wang Z et al (2021) Mutation N501Y in RBD of Spike Protein Strengthens the Interaction between COVID-19 and its Receptor ACE2. https://doi.org/10.1101/2021.02.14.431117. %J bioRxiv

166. Liu Y, Liu J, Plante KS, Plante JA, Xie X, Zhang X et al (2021a) The N501Y spike substitution enhances SARS-CoV-2 transmission. https://doi.org/10.1101/2021.03.08.434499. %J bioRxiv

167. Luan B, Wang H, Huynh T (2021) Molecular Mechanism of the N501Y Mutation for enhanced binding between SARS-CoV-2's Spike protein and human ACE2 Receptor. https://doi.org/10.1101/2021.01.04.425316. %J bioRxiv

168. Natarajan MA, S. Javali P, Jeyaraj Pandian C, Akhtar Ali M, Srivastava V, Jeyaraman J (2021) Computational investigation of increased virulence and pathogenesis of SARS-CoV-2 lineage B.1.1.7. https://doi.org/10.1101/2021.01.25.428190. %J bioRxiv

169. Santos JC, Passos GA (2021) The high infectivity of SARS-CoV-2 B.1.1.7 is associated with increased interaction force between Spike-ACE2 caused by the viral N501Y mutation. https://doi.org/10.1101/2020.12.29.424708. %J bioRxiv

170. Ahmed W, Phillip AM, Biswas KH (2021a) Stable interaction of the UK B.1.1.7 lineage SARS-CoV-2 S1 Spike N501Y mutant with ACE2 revealed by molecular dynamics simulation. https://doi.org/10.1101/2021.01.07.425307. %J bioRxiv

171. Castro A, Carter H, Zanetti M (2021) Potential global impact of the N501Y mutation on MHC-II presentation and immune escape. https://doi.org/10.1101/2021.02.02.429431. %J bioRxiv

172. Richard M, Kok A, de Meulder D, Bestebroer TM, Lamers MM, Okba NMA et al (2020b) SARS-CoV-2 is transmitted via contact and via the air between ferrets. Nat Commun 11(1):3496. https://doi.org/10.1038/s41467-020-17367-2

173. Sun S, Gu H, Cao L, Chen Q, Yang G, Li R-T et al (2020) Characterization and structural basis of a lethal mouse-adapted SARS-CoV-2. https://doi.org/10.1101/2020.11.10.377333. %J bioRxiv

174. Rathnasinghe R, Jangra S, Cupic A, Martinez-Romero C, Mulder LCF, Kehrer T et al (2021) The N501Y mutation in SARS-CoV-2 spike leads to morbidity in obese and aged mice and is neutralized by convalescent and post-vaccination human sera. https://doi.org/10.1101/2021.01.19.21249592 %J medRxiv.

175. Gu H, Chen Q, Yang G, He L, Fan H, Deng Y-Q et al (2020) Adaptation of SARS-CoV-2 in BALB/c mice for testing vaccine efficacy. 369(6511):1603–1607. https://doi.org/10.1126/science.abc4730 %J Science

176. Lassaunière R, Fonager JR, M, Frische A, Strandh C, Rasmussen T, Bøtner A et al (2020a) Working paper on SARS-CoV-2 spike mutations arising in Danish mink, their spread to humans and neutralization data

177. Oude Munnink BB, Sikkema RS, Nieuwenhuijse DF, Molenaar RJ, Munger E, Molenkamp R et al (2020) Transmission of SARS-CoV-2 on mink farms between humans and mink and back to humans. https://doi.org/10.1126/science.abe5901 %J Science

178. Cai HY, Cai A (2021) SARS-CoV-2 spike protein gene variants with N501T and G142D mutation dominated infections in minks in the US. https://doi.org/10.1101/2021.03.18.21253734 %J medRxiv

179. Bedotto M, Fournier P-E, Houhamdi L, Colson P, Raoult D (2021) Implementation of an in-house real-time reverse transcription-PCR assay to detect the emerging SARS-CoV-2 N501Y variants. https://doi.org/10.1101/2021.02.03.21250661 %J medRxiv

180. Ahmed SM, Juvvadi SR, Kalapala R, Sreemanthula JB (2021b) Detection of SARS-CoV-2 N501Y mutation by RT-PCR to identify the UK and the South African strains in the population of South Indian state of Telangana. https://doi.org/10.1101/2021.03.27.21254107 %J medRxiv

181. Abdulnoor M, Eshaghi A, Perusini SJ, Corbeil A, Cronin K, Fittipaldi N et al (2021) Real-time RT-PCR allelic discrimination assay for detection of N501Y mutation in the Spike protein of SARS-CoV-2 associated with variants of concern. https://doi.org/10.1101/2021.06.23.21258782 %J medRxiv.

182. Hayashi T, Yaegashi N, Konishi I (2020) Effect of RBD mutation (Y453F) in spike glycoprotein of SARS-CoV-2 on neutralizing antibody affinity. https://doi.org/10.1101/2020.11.27.401893. %J bioRxiv

183. Lassaunière R, Fonager J, Rasmussen M, Frische A, Polacek Strandh C, Bruun Rasmussen T et al (2020b) Working paper on SARS-CoV-2 spike mutations arising in Danish mink, their spread to humans and neutralization data. https://files.ssi.dk/Mink-cluster-5-short-report_AFO2

184. Yao W, Wang Y, Ma D, Tang X, Wang H, Li C et al (2021) Spike mutations decrease SARS-CoV-2 sensitivity to neutralizing antibodies but not ACE2-Ig in vitro. https://doi.org/10.1101/2021.01.27.428353. %J bioRxiv

185. Hoffmann M, Zhang L, Krüger N, Graichen L, Kleine-Weber H, Hofmann-Winkler H et al (2021b) SARS-CoV-2 mutations acquired in mink reduce antibody-mediated neutralization. https://doi.org/10.1101/2021.02.12.430998. %J bioRxiv

186. Liu Z, VanBlargan LA, Rothlauf PW, Bloyet L-M, Chen RE, Stumpf S et al (2020b) Landscape analysis of escape variants identifies SARS-CoV-2 spike mutations that attenuate monoclonal and serum antibody neutralization. https://doi.org/10.1101/2020.11.06.372037. %J bioRxiv

187. Sabino EC, Buss LF, Carvalho MPS, Prete CA Jr, Crispim MAE, Fraiji NA et al (2021) Resurgence of COVID-19 in Manaus, Brazil, despite high seroprevalence. Lancet 397(10273):452–455. https://doi.org/10.1016/S0140-6736(21)00183-5

188. Di Giacomo S, Mercatelli D, Rakhimov A, Giorgi FM. Preliminary report on severe acute respiratory syndrome coronavirus 2 (SARS-CoV-2) Spike mutation T478K.n/a(n/a). https://doi.org/10.1002/jmv.27062

189. Di Giacomo S, Mercatelli D, Rakhimov A, Giorgi FM (2021) Preliminary report on SARS-CoV-2 Spike mutation T478K. https://doi.org/10.1101/2021.03.28.437369. %J bioRxiv
190. Laiton-Donato K, Usme-Ciro JA, Franco-Munoz C, Alvarez-Diaz DA, Ruiz-Moreno HA, Reales-Gonzalez J et al (2021a) Novel highly divergent SARS-CoV-2 lineage with the Spike substitutions L249S and E484K. https://doi.org/10.1101/2021.03.12.21253000 %J medRxiv.
191. Lesho E, Corey B, Lebreton F, Ong A, Swierczewski B, Bennett J et al (2021) Emergence of the E484K Mutation in SARS-CoV-2 Lineage B.1.1.220 in Upstate New York. https://doi.org/10.1101/2021.03.11.21253231 %J medRxiv.
192. Skidmore PT, Kaelin EA, Holland LA, Maqsood R, Wu LI, Mellor NJ et al (2021) Emergence of a SARS-CoV-2 E484K variant of interest in Arizona. https://doi.org/10.1101/2021.03.26.21254367 %J medRxiv.
193. Hirotsu Y, Omata M (2021) Household transmission of SARS-CoV-2 R.1 lineage with spike E484K mutation in Japan. https://doi.org/10.1101/2021.03.16.21253248 %J medRxiv.
194. Wang WB, Liang Y, Jin YQ, Zhang J, Su JG, Li QM (2021) E484K mutation in SARS-CoV-2 RBD enhances binding affinity with hACE2 but reduces interactions with neutralizing antibodies and nanobodies: binding free energy calculation studies. https://doi.org/10.1101/2021.02.17.431566. %J bioRxiv
195. Greaney AJ, Starr TN, Barnes CO, Weisblum Y, Schmidt F, Caskey M et al (2021b) Mutational escape from the polyclonal antibody response to SARS-CoV-2 infection is largely shaped by a single class of antibodies. https://doi.org/10.1101/2021.03.17.435863. %J bioRxiv
196. Gobeil S, Janowska K, McDowell S, Mansouri K, Parks R, Stalls V et al (2021) Effect of natural mutations of SARS-CoV-2 on spike structure, conformation and antigenicity. https://doi.org/10.1101/2021.03.11.435037. %J bioRxiv
197. Jangra S, Ye C, Rathnasinghe R, Stadlbauer D, Krammer F, Simon V et al (2021) The E484K mutation in the SARS-CoV-2 spike protein reduces but does not abolish neutralizing activity of human convalescent and post-vaccination sera. https://doi.org/10.1101/2021.01.26.21250543 %J medRxiv.
198. Truong TT, Ryutov A, Pandey U, Yee R, Goldberg L, Bhojwani D et al (2021) Persistent SARS-CoV-2 infection and increasing viral variants in children and young adults with impaired humoral immunity. https://doi.org/10.1101/2021.02.27.21252099 %J medRxiv.
199. Weekley CM, Purcell DFJ, Parker MW (2021) SARS-CoV-2 Spike receptor-binding domain with a G485R mutation in complex with human ACE2. https://doi.org/10.1101/2021.03.16.434488. %J bioRxiv
200. Focosi D, Novazzi F, Genoni A, Dentali F, Dalla Gasperina D, Baj AMF (2021) Emergence of SARS-CoV-2 Spike Escape Mutation Q493R After Treatment for COVID-19. Emerging Infect Diseases 27(10). https://doi.org/10.3201/eid2710.211538
201. Guigon A, Faure E, Lemaire C, Chopin M, Tinez C, Assaf A et al (2021) Emergence of Q493R mutation in SARS-CoV-2 spike protein during bamlanivimab/etesevimab treatment and resistance to viral clearance. Submitted
202. Truffot A, Andreani J, Le Marechal M, Caporossi A, Epaulard O, Poignard P et al (2021) Selection and compartmentalization of SARS-Cov2 variants under antibody therapy in an immunocompromised patient. Emerging infectious diseases
203. Rubsamen R, Burkholz S, Pokhrel S, Kraemer BR, Mochly-Rosen D, Carback RT et al (2020) Paired SARS CoV-2 Spike protein mutations observed during ongoing SARS-CoV-2 viral transfer from humans to minks and back to humans. https://doi.org/10.1101/2020.12.22.424003. %J bioRxiv
204. Control ECfDPa (2020) Rapid risk assessment: detection of new SARS-CoV-2 variants related to mink. https://www.ecdc.europa.eu/en/publications-data/detection-new-sars-cov-2-variants-mink. Accessed 12 Nov
205. Lu L, Sikkema RS, Velkers FC, Nieuwenhuijse DF, Fischer EAJ, Meijer PA et al (2021a) Adaptation, spread and transmission of SARS-CoV-2 in farmed minks and related humans in the Netherlands. https://doi.org/10.1101/2021.07.13.452160. %J bioRxiv

206. Larsen HD, Fonager J, Lomholt FK, Dalby T, Benedetti G, Kristensen B et al (2020) Preliminary report of an outbreak of SARS-CoV-2 in mink and mink farmers associated with community spread, Denmark, June to November 2020. 26(5):2100009. https://doi.org/10.2807/1560-7917.ES.2021.26.5.210009.

207. Rabalski L, Kosinski M, Smura T, Aaltonen K, Kant R, Sironen T et al (2020) Detection and molecular characterisation of SARS-CoV-2 in farmed mink (Neovision vision) in Poland. https://doi.org/10.1101/2020.12.24.422670. %J bioRxiv

208. Rabalski L, Kosinski M, Mazur-Panasiuk N, Szewczyk B, Bienkowska-Szewczyk K, Kant R et al (2021) Zoonotic spillover of SARS-CoV-2: mink-adapted virus in humans. https://doi.org/10.1101/2021.03.05.433713. %J bioRxiv

209. Konishi T (2020) SARS-CoV-2 mutations among minks show reduced lethality and infectivity to humans. https://doi.org/10.1101/2020.12.23.424267. %J bioRxiv

210. Bayarri-Olmos R, Rosbjerg A, Johnsen LB, Helgstrand C, Bak-Thomsen T, Garred P et al (2021) The SARS-CoV-2 Y453F mink variant displays a striking increase in ACE-2 affinity but does not challenge antibody neutralization. https://doi.org/10.1101/2021.01.29.428834. %J bioRxiv

211. Eckstrand C, Baldwin T, Torchetti MK, Killian ML, Rood KA, Clayton M et al (2021) An outbreak of SARS-CoV-2 with high mortality in mink (Neovison vison) on multiple Utah farms. https://doi.org/10.1101/2021.06.09.447754. %J bioRxiv

212. Leung K, Shum MH, Leung GM, Lam TT, Wu JT (2020) Early empirical assessment of the N501Y mutant strains of SARS-CoV-2 in the United Kingdom, October to November 2020. https://www.medrxiv.org/content/medrxiv/early/2020/12/22/2020.12.20.20248581.full.pdf

213. Davies NG, Barnard RC, Jarvis CI, Kucharski AJ, Munday JD, Pearson CAB et al (2020) Estimated transmissibility and severity of novel SARS-CoV-2 variant of concern 202012/01 in England. https://doi.org/10.1101/2020.12.24.20248822 %J medRxiv

214. Vöhringer H, Sinnott M, Amato R, Martincorena I, Kwiatkowski D, Barrett J, et al (2020) Lineage-specific growth of SARS-CoV-2 B.1.1.7 during the English national lockdown. https://virological.org/t/lineage-specific-growth-of-sars-cov-2-b-1-1-7-during-the-english-national-lockdown/575. Accessed 31 Dec 2020

215. Lindstrøm JC, Engebretsen S, Kristoffersen AB, Rø GØI, Palomares AD-L, Engø-Monsen K et al (2021) Increased transmissibility of the B.1.1.7 SARS-CoV-2 variant: Evidence from contact tracing data in Oslo, January to February 2021. https://doi.org/10.1101/2021.03.29.21254122 %J medRxiv

216. van Dorp CH, Goldberg EE, Hengartner N, Ke R, Romero-Severson EO (2021) Estimating the strength of selection for new SARS-CoV-2 variants. https://doi.org/10.1101/2021.03.29.21254233 %J medRxiv

217. Lyngse FP, Mølbak K, Skov RL, Christiansen LE, Mortensen LH, Albertsen MP et al (2021) Increased transmissibility of SARS-CoV-2 lineage B.1.1.7 by age and viral load: evidence from Danish households. https://doi.org/10.1101/2021.04.16.21255459 %J medRxiv

218. Meister TL, Fortmann J, Todt D, Heinen N, Ludwig A, Brueggemann Y et al (2021) Comparable environmental stability and disinfection profiles of the currently circulating SARS-CoV-2 variants of concern B.1.1.7 and B.1.351. https://doi.org/10.1101/2021.04.07.438820. %J bioRxiv

219. Hassan SS, Kodakandla V, Redwan EM, Lundstrom K, Pal Choudhury P, Abd El-Aziz TM et al (2021) An issue of concern: unique truncated ORF8 protein variants of SARS-CoV-2. https://doi.org/10.1101/2021.05.25.445557. %J bioRxiv

220. Yang J, Zhang G, Yu D, Cao R, Wu X, Ling Y et al (2021c) Evolutionary insights into a non-coding deletion of SARS-CoV-2 B.1.1.7. https://doi.org/10.1101/2021.04.30.442029. %J bioRxiv

221. Kuo C-W, Yang T-J, Chien Y-C, Yu P-Y, Hsu S-TD, Khoo K-H (2021) Distinct shifts in site-specific glycosylation pattern of SARS-CoV-2 spike proteins associated with arising mutations in the D614G and Alpha variants. https://doi.org/10.1101/2021.07.21.453140. %J bioRxiv

222. Thorne LG, Bouhaddou M, Reuschl A-K, Zuliani-Alvarez L, Polacco BJ, Pelin A et al (2021) Evolution of enhanced innate immune evasion by the SARS-CoV-2 B.1.1.7 UK variant. https://doi.org/10.1101/2021.06.06.446826. %J bioRxiv

223. England PH (2021) Investigation of novel SARS-CoV-2 variant of concern 202012/01. Techn Brief 5. https://assets.publishing.service.gov.uk/government/uploads/system/uploads/attach ment_data/file/957504/Variant_of_Concern_VOC_202012_01_Technical_Briefing_5_Engl and.pdf. Accessed 2 Feb 2021

224. Carreño JM, Alshammary H, Singh G, Raskin A, Amanat F, Amoako A et al (2021) Reduced neutralizing activity of post-SARS-CoV-2 vaccination serum against variants B.1.617.2, B.1.351, B.1.1.7+E484K and a sub-variant of C.37. 2021:2021.07.21.21260961. https://doi. org/10.1101/2021.07.21.21260961 %J medRxiv

225. Liu Y, Liu J, Xia H, Zhang X, Zou J, Fontes-Garfias CR et al (2021) BNT162b2-Elicited Neutralization against New SARS-CoV-2 Spike Variants. 385(5):472–474. https://doi.org/ 10.1056/NEJMc2106083

226. Parker MD, Lindsey BB, Shah DR, Hsu S, Keeley AJ, Partridge DG et al (2021) Altered subge-nomic RNA expression in SARS-CoV-2 B.1.1.7 infections. https://doi.org/10.1101/2021.03. 02.433156. %J bioRxiv

227. Wollschlaeger P, Gerlitz N, Todt D, Pfaender S, Bollinger T, Sing A et al (2021) SARS-CoV-2 N gene dropout and N gene Ct value shift as indicator for the presence of B.1.1.7 lineage in a widely used commercial multiplex PCR assay. https://doi.org/10.1101/2021.03.23.21254171 %J medRxiv.

228. Lopez-Rincon A, Perez-Romero C, Tonda A, Mendoza-Maldonado L, Claassen E, Garssen J et al (2020) Design of Specific Primer Set for Detection of B.1.1.7 SARS-CoV-2 Variant using deep learning. https://www.biorxiv.org/content/biorxiv/early/2020/12/29/2020.12.29. 424715.full.pdf

229. Lopez-Rincon A, Perez-Romero C, Tonda A, Mendoza-Maldonado L, Claassen E, Garssen J, et al (2021) Design of specific primer sets for the detection of B.1.1.7, B.1.351 and P.1 SARS-CoV-2 variants using deep learning. https://www.biorxiv.org/content/biorxiv/early/2021/01/ 21/2021.01.20.427043.full.pdf. Accessed 27 Jan 2021

230. Kovacova V, Boršová KP, Radvanszka EDM, Hajdu R, Čabanová VS, Ličková M, Lukáčiková L et al (2020) A novel, room temperature-stable, multiplexed RT-qPCR assay to distinguish lineage B.1.1.7 from the remaining SARS-CoV-2 lineages.https://virological.org/t/a-novel-room-temperature-stable-multiplexed-rt-qpcr-assay-to-distinguish-lineage-b-1-1-7-from-the-remaining-sars-cov-2-lineages/611

231. Perchetti GA, Zhu H, Mills MG, Shrestha L, Wagner C, Bakhash SM et al (2021) Specific allelic discrimination of N501Y and other SARS-CoV-2 mutations by ddPCR detects B.1.1.7 lineage in Washington State. https://doi.org/10.1101/2021.03.10.21253321 %J medRxiv

232. Zhang J, Zhang Y, Kang J, Chen S, He Y, Han B et al (2021d) Potential transmission chains of variant B.1.1.7 and co-mutations of SARS-CoV-2. https://doi.org/10.1101/2021.04.16. 440141. %J bioRxiv

233. Xie X, Lewis T-J, Green N, Wang Z (2021a) Phylogenetic network analysis revealed the recombinant origin of the SARS-CoV-2 VOC202012/01 (B.1.1.7) variant first discovered in U.K. https://doi.org/10.1101/2021.06.24.449840. %J bioRxiv

234. Jackson B, Rambaut A, Pybus O, Robertson D, Connor T, Loman N et al (2021b) Recombinant SARS-CoV-2 genomes involving lineage B.1.1.7 in the UK. https://virological.org/t/recomb inant-sars-cov-2-genomes-involving-lineage-b-1-1-7-in-the-uk/658 (2021).

235. Khavari P, Ramanathan M, Ferguson ID, Miao W (2021) SARS-CoV-2 B.1.1.7 and B.1.351 Spike variants bind human ACE2 with increased affinity. https://doi.org/10.1101/2021.02.22. 432359. %J bioRxiv

236. Fiedler S, Denninger V, Morgunov AS, Ilsley A, Worth R, Meisl G et al (2021) Mutations in two SARS-CoV-2 variants of concern reflect two distinct strategies of antibody escape. https://doi.org/10.1101/2021.07.23.453327. %J bioRxiv

237. Rajah MM, Hubert M, Bishop E, Saunders N, Robinot R, Grzelak L et al (2021) B.1.1.7 and B.1.351 SARS-CoV-2 variants display enhanced Spike-mediated fusion. https://doi.org/10. 1101/2021.06.11.448011. %J bioRxiv

238. Grabowski F, Preibisch G, Kochanczyk M, Lipniacki T (2021b) SARS−CoV−2 Variant under investigation 202012/01 has more than twofold replicative advantage. https://doi.org/10.1101/ 2020.12.28.20248906. %J medRxiv

239. Brown JC, Goldhill DH, Zhou J, Peacock TP, Frise R, Goonawardane N et al (2021) Increased transmission of SARS-CoV-2 lineage B.1.1.7 (VOC 2020212/01) is not accounted for by a replicative advantage in primary airway cells or antibody escape. https://doi.org/10.1101/2021.02.24.432576. %J bioRxiv

240. Prince T, Dong X, Penrice-Randal R, Randle N, Hartley C, Goldswain H et al (2021) Sequence analysis of SARS-CoV-2 in nasopharyngeal samples from patients with COVID-19 illustrates population variation and diverse phenotypes, placing the in vitro growth properties of B.1.1.7 and B.1.351 lineage viruses in context. https://doi.org/10.1101/2021.03.30.437704. %J bioRxiv

241. Touret F, Luciani L, Baronti C, Cochin M, Driouich J-S, Gilles M et al (2021) Replicative fitness SARS-CoV-2 20I/501Y.V1 variant in a human reconstituted bronchial epithelium. https://doi.org/10.1101/2021.03.22.436427. %J bioRxiv

242. Dicken SJ, Murray MJ, Thorne LG, Reuschl A-K, Forrest C, Ganeshalingham M et al (2021) Characterisation of B.1.1.7 and Pangolin coronavirus spike provides insights on the evolutionary trajectory of SARS-CoV-2. https://doi.org/10.1101/2021.03.22.436468. %J bioRxiv

243. Lamers MM, Breugem TI, Mykytyn AZ, Wang Y, Groen N, Knoops K et al (2021) Human organoid systems reveal in vitro correlates of fitness for SARS-CoV-2 B.1.1.7. https://doi.org/10.1101/2021.05.03.441080. %J bioRxiv

244. Mohandas S, Yadav PD, Nyayanit D, Deshpande G, Aich AS, Sapkal G et al (2021a) Comparison of the pathogenicity and virus shedding of SARS CoV-2 VOC 202012/01 and D614G variant in hamster model. https://doi.org/10.1101/2021.02.25.432136. %J bioRxiv

245. Cochin M, Luciani L, Touret F, Driouich JS, Petit P-R, Moureau G et al (2021) SARS-CoV-2 20I/501Y.V1 variant in the hamster model. https://doi.org/10.1101/2021.04.19.440435. %J bioRxiv

246. Port J, Yinda KC, Avanzato V, Schulz J, Holbrook M, van Doremalen N et al (2021) Increased aerosol transmission for B.1.1.7 (alpha variant) over lineage A variant of SARS-CoV-2. https://doi.org/10.1101/2021.07.26.453518. %J bioRxiv

247. Abdelnabi R, Boudewijns R, Foo CS-Y, Seldeslachts L, Sanchez-Felipe L, Zhang X et al (2021) Comparative infectivity and pathogenesis of emerging SARS-CoV-2 variants in Syrian hamsters. https://doi.org/10.1101/2021.02.26.433062. %J bioRxiv

248. Nunez IA, Lien CZ, Selvaraj P, Stauft CB, Liu S, Starost M et al (2021) SARS-CoV-2 B.1.1.7 infection of Syrian hamster does not cause more severe disease and is protected by naturally acquired immunity. https://doi.org/10.1101/2021.04.02.438186. %J bioRxiv

249. Radvak P, Kwon HJ, Kosikova M, Ortega-Rodriguez U, Xiang R, Phue J-N et al (2021) B.1.1.7 and B.1.351 variants are highly virulent in K18-ACE2 transgenic mice and show different pathogenic patterns from early SARS-CoV-2 strains. https://doi.org/10.1101/2021.06.05.447221. %J bioRxiv

250. Volz E, Mishra S, Chand M, Barrett J, Johnson R, Geidelberg L et al (2020) Transmission of SARS-CoV-2 Lineage B.1.1.7 in England: Insights from linking epidemiological and genetic data. https://virological.org/t/transmission-of-sars-cov-2-lineage-b-1-1-7-in-england-insights-from-linking-epidemiological-and-genetic-data/576. Accessed 31 Dec 2020

251. Graham MS, Sudre CH, May A, Antonelli M, Murray B, Varsavsky T et al (2021a) The effect of SARS-CoV-2 variant B.1.1.7 on symptomatology, re-infection and transmissibility. https://doi.org/10.1101/2021.01.28.21250680 %J medRxiv

252. Davies NG, Jarvis CI, Edmunds WJ, Jewell NP, Diaz-Ordaz K, Keogh RH (2021) Increased hazard of death in community-tested cases of SARS-CoV-2 Variant of Concern 202012/01. https://doi.org/10.1101/2021.02.01.21250959. %J medRxiv

253. Jablonska K, Aballea S, Auquier P, Toumi M. On the association between SARS-COV-2 variants and COVID-19 mortality during the second wave of the pandemic in Europe. https://doi.org/10.1101/2021.03.25.21254289. %J medRxiv

254. Patone M, Thomas K, Hatch R, Tan PS, Liao W, Coupland C et al (2021) Analysis of severe outcomes associated with the SARS-CoV-2 Variant of Concern 202012/01 in England using ICNARC Case Mix Programme and QResearch databases. https://doi.org/10.1101/2021.03.11.21253364. %J medRxiv

255. Snell LB, Wang W, Alcolea-Medina A, Charalampous T, Nebbia G, Batra R et al (2021) First and second SARS-CoV-2 waves in inner London: A comparison of admission characteristics and the effects of the B.1.1.7 variant. https://doi.org/10.1101/2021.03.16.21253377. %J medRxiv

256. Xiao C, Mao L, Wang Z, Zhu G, Gao L, Su J et al (2021) SARS-CoV-2 variant B.1.1.7 caused HLA-A2+ CD8+ T cell epitope mutations for impaired cellular immune response. https://doi.org/10.1101/2021.03.28.437363. %J bioRxiv

257. Xie X, Zou J, Fontes-Garfias CR, Xia H, Swanson KA, Cutler M et al (2021b) Neutralization of N501Y mutant SARS-CoV-2 by BNT162b2 vaccine-elicited sera. https://www.biorxiv.org/content/biorxiv/early/2021/01/07/2021.01.07.425740.full.pdf

258. Collier D, Meng B, Ferreira I, Datir R, Temperton NJ, Elmer A et al (2021) Impact of SARS-CoV-2 B.1.1.7 Spike variant on neutralisation potency of sera from individuals vaccinated with Pfizer vaccine BNT162b2. https://doi.org/10.1101/2021.01.19.21249840. %J medRxiv

259. Xie X, Liu Y, Liu J, Zhang X, Zou J, Fontes-Garfias CR et al (2021c) Neutralization of spike 69/70 deletion, E484K, and N501Y SARS-CoV-2 by BNT162b2 vaccine-elicited sera. https://www.biorxiv.org/content/biorxiv/early/2021/01/27/2021.01.27.427998.full.pdf

260. Tada T, Dcosta BM, Samanovic-Golden M, Herati RS, Cornelius A, Mulligan MJ et al (2021b) Neutralization of viruses with European, South African, and United States SARS-CoV-2 variant spike proteins by convalescent sera and BNT162b2 mRNA vaccine-elicited antibodies. https://doi.org/10.1101/2021.02.05.430003. %J bioRxiv

261. Muik A, Wallisch A-K, Saenger B, Swanson KA, Muehl J, Chen W et al (2021) Neutralization of SARS-CoV-2 lineage B.1.1.7 pseudovirus by BNT162b2 vaccine-elicited human sera. https://doi.org/10.1101/2021.01.18.426984. %J bioRxiv

262. Planas D, Bruel T, Grzelak L, Guivel-Benhassine F, Staropoli I, Porrot F et al (2021a) Sensitivity of infectious SARS-CoV-2 B.1.1.7 and B.1.351 variants to neutralizing antibodies. https://doi.org/10.1101/2021.02.12.430472. %J bioRxiv

263. Kuzmina A, Khalaila Y, Voloshin O, Keren-Naus A, Bohehm L, Raviv Y et al (2021) SARS CoV-2 escape variants exhibit differential infectivity and neutralization sensitivity to convalescent or post-vaccination sera. https://doi.org/10.1101/2021.02.22.21252002. %J medRxiv

264. Widera M, Wilhelm A, Hoehl S, Pallas C, Kohmer N, Wolf T et al (2021) Bamlanivimab does not neutralize two SARS-CoV-2 variants carrying E484K in vitro. https://doi.org/10.1101/2021.02.24.21252372. %J medRxiv

265. Marot SS, Malet I, Jary A, Leducq V, Abdi B, Teyssou E et al (2021) Neutralization heterogeneity of United Kingdom and South-African SARS CoV-2 variants in BNT162b2-vaccinated or convalescent COVID-19 healthcare workers. https://doi.org/10.1101/2021.03.05.434089. %J bioRxiv

266. Wu K, Werner AP, Moliva JI, Koch M, Choi A, Stewart-Jones GBE et al (2021b) Serum neutralizing activity elicited by mRNA-1273 vaccine. New England J Med. https://doi.org/10.1056/NEJMc2102179

267. Wang P, Nair MS, Lihong L, Iketani S, Luo Y, Guo Y et al (2021) Antibody resistance of SARS-CoV-2 variants B.1.351 and B.1.1.7. Nature 593:130–135. https://doi.org/10.1038/s41586-021-03398-2

268. Edara VV, Floyd K, Lai L, Gardner M, Hudson W, Piantadosi A et al (2021a) Infection and mRNA-1273 vaccine antibodies neutralize SARS-CoV-2 UK variant. https://doi.org/10.1101/2021.02.02.21250799. %J medRxiv

269. Sapkal GN, Yadav P, Ella R, Deshpande G, Sahay R, Gupta N et al (2021a) Neutralization of UK-variant VUI-202012/01 with COVAXIN vaccinated human serum. https://doi.org/10.1101/2021.01.26.426986. %J bioRxiv

270. Lumley SF, Rodger G, Constantinides B, Sanderson N, Chau KK, Street TL et al (2021) An observational cohort study on the incidence of SARS-CoV-2 infection and B.1.1.7 variant infection in healthcare workers by antibody and vaccination status. https://doi.org/10.1101/2021.03.09.21253218 %J medRxiv.

271. Graham C, Seow J, Huettner I, Khan H, Kouphou N, Acors S et al (2021) Impact of the B.1.1.7 variant on neutralizing monoclonal antibodies recognizing diverse epitopes on SARS–CoV–2 Spike. https://doi.org/10.1101/2021.02.03.429355. %J bioRxiv

272. Cerutti G, Rapp M, Guo Y, Bahna F, Bimela J, Reddem ER et al (2021) Structural basis for accommodation of emerging B.1.351 and B.1.1.7 variants by two potent SARS-CoV-2 neutralizing antibodies. https://doi.org/10.1101/2021.02.21.432168. %J bioRxiv

273. Clark JJ, Sharma P, Bentley E, Harding AC, Kipar A, Neary M et al (2021) Naturally-acquired immunity in Syrian Golden Hamsters provides protection from re-exposure to emerging heterosubtypic SARS-CoV-2 variants B.1.1.7 and B.1.351. https://doi.org/10.1101/2021.03. 10.434447. %J bioRxiv

274. Harrington D, Kele B, Pereira S, Couto-Parada X, Riddell A, Forbes S et al (2021) Confirmed reinfection with SARS-CoV-2 variant VOC-202012/01. Clin Infect Dis. https://doi.org/10. 1093/cid/ciab014

275. Novazzi F, Baj A, Genoni A, Spezia PG, Colombo A, Cassani G et al (2021) SARS-CoV-2 B.1.1.7 reinfection after previous COVID-19 in two immunocompetent Italian patients. J Med Virol https://doi.org/10.1002/jmv.27066

276. Marquez L, Koy T, Spinler JK, Luna RA, Tocco L, Fasciano L et al(2021) Reinfection with severe acute respiratory syndrome coronavirus 2 (SARS-CoV-2) B.1.1.7 variant in an immuno-compromised adolescent. Infect Control Hospital Epidemiol 1–2. https://doi.org/10.1017/ice. 2021.195

277. Vankeerberghen A, Holderberke A, Boel A, Van Vaerenbergh K, De Beehouwer H, Cattoir L (2021) Case report: a 90 year-old lady infected withb two COVID19 VoCs: 20I/501Y.V1 and 20H/501Y.V2. ECCMID2021. p. Abstract 04978

278. Haynes WA, Kamath K, Lucas C, Shon J, Iwasaki A (2021) Impact of B.1.1.7 variant mutations on antibody recognition of linear SARS-CoV-2 epitopes. https://doi.org/10.1101/2021.01.06. 20248960. %J medRxiv

279. Faulkner N, Ng K, Wu M, Harvey R, Hussain S, Greco M et al (2021) Reduced antibody cross-reactivity following infection with B.1.1.7 than with parental SARS-CoV-2 strains. https:// doi.org/10.1101/2021.03.01.433314. %J bioRxiv

280. McCallum M, Bassi J, De Marco A, Chen A, Walls AC, Di Iulio J et al (2021) SARS-CoV-2 immune evasion by variant B.1.427/B.1.429. Science https://doi.org/10.1126/science.abi7994

281. Rambaut A, Loman N, Pybus O, Barclay W, Barrett J, Carabelli A et al (2020b) Preliminary genomic characterisation of an emergent SARS-CoV-2 lineage in the UK defined by a novel set of spike mutations. https://virological.org/t/preliminary-genomic-characterisation-of-an-emergent-sars-cov-2-lineage-in-the-uk-defined-by-a-novel-set-of-spike-mutations/563

282. Gozzi N, Chinazzi M, Davis JT, Mu K, Pastore y Piontti A, Ajelli M et al (2021) Estimating the spreading and dominance of SARS-CoV-2 VOC 202012/01 (lineage B.1.1.7) across Europe. https://doi.org/10.1101/2021.02.22.21252235. %J medRxiv

283. Du Z, Wang L, Yang B, Ali ST, Yang B, K. Tsang T et al (2021) International risk of the new variant COVID-19 importations originating in the United Kingdom. https://doi.org/10.1101/ 2021.01.09.21249384. %J medRxiv

284. O'Toole A, Hill V, Pybus O, Watts A, Bogoch I, Khan K et al (2021) Tracking the international spread of SARS-CoV-2 lineages B.1.1.7 and B.1.351/501Y-V2. https://virological.org/t/tra cking-the-international-spread-of-sars-cov-2-lineages-b-1-1-7-and-b-1-351-501y-v2/592

285. Vrancken B, Dellicour S, Smith DM, Chaillon A (2021) Phylogenetic analyses of SARS-CoV-2 B.1.1.7 lineage suggest a single origin followed by multiple exportation events versus convergent evolution. https://doi.org/10.1101/2021.01.12.426373. %J bioRxiv

286. Smith TP, Dorigatti I, Mishra S, Volz E, Walker PGT, Ragonnet-Cronin M et al (2021) Environ-mental drivers of SARS-CoV-2 lineage B.1.1.7 transmission in England, October to December 2020. https://doi.org/10.1101/2021.03.09.21253242. %J medRxiv.

287. Kosakovsky Pond S, Wilkison E, Weaver S, James S, Tegally H, de Oliveira T et al (2020) A preliminary selection analysis of the South African V501.V2 SARS-CoV-2 clade. https://virological.org/t/a-preliminary-selection-analysis-of-the-south-african-v501-v2-sars-cov-2-clade/573. Accessed 31 Dec 2020

288. Zhang Y, Tanner NA (2021) Improving RT-LAMP detection of SARS-CoV-2 RNA through primer set selection and combination. https://doi.org/10.1101/2021.07.23.453545. %J bioRxiv

289. Teyssou E, Soulie C, Visseaux B, Lambert-Niclot S, Ferre V, Marot S et al (2021) The 501Y.V2 SARS-CoV-2 variant has an intermediate viral load between the 501Y.V1 and the historical variants in nasopharyngeal samples from newly diagnosed COVID-19 patients. https://doi.org/10.1101/2021.03.21.21253498. %J medRxiv.

290. Hoffmann M, Arora P, Gross R, Seidel A, Hoernich B, Hahn A et al (2021) SARS-CoV-2 variants B.1.351 and P.1 escape from neutralizing antibodies. Cell. 184(9):2384–2393. https://doi.org/10.1101/2021.02.11.430787. %J bioRxiv

291. Cheng MH, Krieger JM, Kaynak B, Arditi MA, Bahar I (2021b) Impact of South African 501.V2 Variant on SARS-CoV-2 Spike infectivity and neutralization: a structure-based computational assessment. https://www.biorxiv.org/content/biorxiv/early/2021/01/11/2021.01.10.426143.full.pdf

292. Zhu X, Mannar D, Srivastava SS, Berezuk A, Demers J-P, Saville J et al (2021) Cryo-EM structure of the N501Y SARS-CoV-2 Spike protein in complex with a potent neutralizing antibody. https://doi.org/10.1101/2021.01.11.426269. %J bioRxiv

293. Liu H, Wei P, Zhang Q, Chen Z, Aviszus K, Downing W et al (2021b) 501Y.V2 and 501Y.V3 variants of SARS-CoV-2 lose binding to Bamlanivimab in vitro. https://doi.org/10.1101/2021.02.16.431305. %J bioRxiv

294. Wibmer CK, Ayres F, Hermanus T, Madzivhandila M, Kgagudi P, Lambson BE et al (2021) SARS-CoV-2 501Y.V2 escapes neutralization by South African COVID-19 donor plasma. Nature medicine. https://doi.org/10.1038/s41591-021-01285-x

295. Edara VV, Norwood C, Floyd K, Lai L, Davis-Gardner ME, Hudson WH et al (2021b) Reduced binding and neutralization of infection- and vaccine-induced antibodies to the B.1.351 (South African) SARS-CoV-2 variant. https://doi.org/10.1101/2021.02.20.432046. %J bioRxiv

296. Cele S, Gazy I, Jackson L, Hwa S-H, Tegally H, Lustig G et al (2021) Escape of SARS-CoV-2 501Y.V2 variants from neutralization by convalescent plasma. Nature. 10.1038-s41586-021-03471-w

297. Riou C, Keeton R, Moyo-Gwete T, Hermanus T, Kgagudi P, Baguma R et al (2021) Loss of recognition of SARS-CoV-2 B.1.351 variant spike epitopes but overall preservation of T cell immunity. https://doi.org/10.1101/2021.06.03.21258307. %J medRxiv

298. Ravi K, Lauri K, Teemu S, Tobias LF, Sawan Kumar J, Kari A et al (2021) Common laboratory mice are susceptible to infection with SARS-CoV2 beta variant. Res Square. https://doi.org/10.21203/rs.3.rs-762916/v1

299. Tarres-Freixas F, Trinite B, Pons-Grifols A, Romero-Durana M, Riveira-Munoz E, Avila-Nieto C et al (2021) SARS-CoV-2 B.1.351 (beta) variant shows enhanced infectivity in K18-hACE2 transgenic mice and expanded tropism to wildtype mice compared to B.1 variant. https://doi.org/10.1101/2021.08.03.454861. %J bioRxiv

300. Abu-Raddad LJ, Chemaitelly H, Ayoub HH, YASSINE HM, Benslimane F, Al Khatib HA et al (2021a) Severity, criticality, and fatality of the SARS-CoV-2 Beta variant. https://doi.org/10.1101/2021.08.02.21261465. %J medRxiv

301. Zucman N, Uhel F, Descamps D, Roux D, Ricard JD (2021) Severe reinfection with South African SARS-CoV-2 variant 501Y.V2: a case report. Clin Infect Dis: Offi Publ Infect Dis Soci Am. https://doi.org/10.1093/cid/ciab129

302. Moyo-Gwete T, Madzivhandila M, Makhado Z, Ayres F, Mhlanga D, Oosthuysen B et al (2021) SARS-CoV-2 501Y.V2 (B.1.351) elicits cross-reactive neutralizing antibodies. https://doi.org/10.1101/2021.03.06.434193. %J bioRxiv

303. Kosakovsky Pond S, Wilkison E, Weaver S, James S, Tegally H, de Oliveira T et al (2021) A preliminary selection analysis of the South African V501.V2 SARS-CoV-2 clade. https://virological.org/t/a-preliminary-selection-analysis-of-the-south-african-v501-v2-sars-cov-2-clade/573

304. Mohsen M, Bachmann MF, Vogel M, Augusto GS, Liu X, Chang X (2021) BNT162b2 mRNA COVID-19 vaccine induces antibodies of broader cross-reactivity than natural infection but

recognition of mutant viruses is up to 10-fold reduced. https://doi.org/10.1101/2021.03.13. 435222. %J bioRxiv

305. Stamatatos L, Czartoski J, Wan Y-H, Homad L, Rubin V, Glantz H et al (2021) Antibodies elicited by SARS-CoV-2 infection and boosted by vaccination neutralize an emerging variant and SARS-CoV-1. medrXiv [Preprint]. https://doi.org/10.1101/2021.02.05.21251182. %J medRxiv

306. Huang B, Dai L, Wang H, Hu Z, Tan W, Gao GF et al (2021) Neutralization of SARS-CoV-2 VOC 501Y.V2 by human antisera elicited by both inactivated BBIBP-CorV and recombinant dimeric RBD ZF2001 vaccines. https://www.biorxiv.org/content/biorxiv/early/2021/02/02/2021.02.01.429069.full.pdf

307. Madhi SA, Baillie V, Cutland CL, Voysey M, Koen AL, Fairlie L et al (2021) Efficacy of the ChAdOx1 nCoV-19 Covid-19 Vaccine against the B.1.351 Variant. New England J Med. https://doi.org/10.1056/NEJMoa2102214

308. Yinda KC, Port J, Bushmaker T, Fischer R, Schulz J, Holbrook M et al (2021) Prior aerosol infection with lineage A SARS-CoV-2 variant protects hamsters from disease, but not reinfection with B.1.351 SARS-CoV-2 variant. https://doi.org/10.1101/2021.05.05.442780. %J bioRxiv

309. Franceschi VB, Ferrareze PAG, Zimerman RA, Cybis GB, Thompson CE. Mutation hotspots, geographical and temporal distribution of SARS-CoV-2 lineages in Brazil, February 2020 to February 2021: insights and limitations from uneven sequencing efforts. 2021:2021.03.08.21253152. https://doi.org/10.1101/2021.03.08.21253152 %J medRxiv.

310. Francisco Junior RdS, Benites LF, Lamarca AP, de Almeida LGP, Hansen AW, Gularte JS et al (2021) Pervasive transmission of E484K and emergence of VUI-NP13L with evidence of SARS-CoV-2 co-infection events by two different lineages in Rio Grande do Sul, Brazil. https://doi.org/10.1101/2021.01.21.21249764. %J medRxiv.

311. Voloch CM, Silva F Rd, de Almeida LGP, Cardoso CC, Brustolini OJ, Gerber AL et al (2020) Genomic characterization of a novel SARS-CoV-2 lineage from Rio de Janeiro, Brazil. https://www.medrxiv.org/content/medrxiv/early/2020/12/26/2020.12.23.202 48598.full.pdf. Accessed 27 Jan 2021

312. Lamarca AP, Almeida LGPd, Francisco RdS, Lima LFA, Scortecci KC, Perez VP et al (2021) Genomic surveillance of SARS-CoV-2 tracks early interstate transmission of P.1 lineage and diversification within P.2 clade in Brazil. https://doi.org/10.1101/2021.03.21.21253418. %J medRxiv

313. Imai M, Halfmann PJ, Yamayoshi S, Iwatsuki-Horimoto K, Chiba S, Watanabe T et alCharacterization of a new SARS-CoV-2 variant that emerged in Brazil. Proc Natl Acad Sci USA 118(27). https://doi.org/10.1073/pnas.2106535118

314. Buss LF, Prete CA Jr, Abrahim CMM, Mendrone A Jr, Salomon T, de Almeida-Neto C et al (2021) Three-quarters attack rate of SARS-CoV-2 in the Brazilian Amazon during a largely unmitigated epidemic. Science 371(6526):288–292. https://doi.org/10.1126/science.abe9728

315. Oliveira MHSd, Lippi G, Henry BM (2021a) Sudden rise in COVID-19 case fatality among young and middle-aged adults in the south of Brazil after identification of the novel B.1.1.28.1 (P.1) SARS-CoV-2 strain: analysis of data from the state of Parana. https://doi.org/10.1101/2021.03.24.21254046. %J medRxiv

316. Coutinho RM, Marquitti FMD, Ferreira LS, Borges ME, Silva RLPd, Canton O et al (2021) Model-based evaluation of transmissibility and reinfection for the P.1 variant of the SARS-CoV-2. https://doi.org/10.1101/2021.03.03.21252706. %J medRxiv

317. He D, Fan G, Wang X, Li Y, Peng Z (2021) The new SARS-CoV-2 variant and reinfection in the resurgence of COVID-19 outbreaks in Manaus, Brazil. https://doi.org/10.1101/2021.03.25.21254281 %J medRxiv.

318. Naveca F, da Costa C, Nascimento V, Souza V, Corado A, Nascimento F, et al (2021a) SARS-CoV-2 reinfection by the new Variant of Concern (VOC) P.1 in Amazonas, Brazil. https://virological.org/t/sars-cov-2-reinfection-by-the-new-variant-of-concern-voc-p-1-in-amazonas-brazil/596. Accessed 27 Jan 2021

319. Prete CA, Buss LF, Abrahim CMM, Salomon T, Crispim MAE, Oikawa MK et al (2021) Reinfection by the SARS-CoV-2 P.1 variant in blood donors in Manaus, Brazil. https://doi.org/10.1101/2021.05.10.21256644. %J medRxiv

320. Martins AF, Zavascki AP, Wink PL, Volpato FCZ, Monteiro FL, Rosset C et al (2021) Detection of SARS-CoV-2 lineage P.1 in patients from a region with exponentially increasing hospitalization rates in February 2021, Rio Grande do Sul, Southern Brazil. https://doi.org/10.1101/2021.03.09.21253204. %J medRxiv

321. Maggi F, Novazzi F, Genoni A, Baj A, Spezia PG, Focosi D et al (2021) Imported SARS-COV-2 variant P.1 detected in traveler returning from Brazil to Italy. Emerging Infect Dis J 27(4). https://doi.org/10.3201/eid2704.210183

322. Matic N, Lowe CF, Ritchie G, Stefanovic A, Lawson T, Jang W et al (2021) Rapid detection of SARS-CoV-2 variants of concern identifying a cluster of B.1.1.28/P.1 variant in British Columbia, Canada. https://doi.org/10.1101/2021.03.04.21252928. %J medRxiv

323. Nelson G, Buzko O, Spilman PR, Niazi K, Rabizadeh S, Soon-Shiong PR (2021) Molecular dynamic simulation reveals E484K mutation enhances spike RBD-ACE2 affinity and the combination of E484K, K417N and N501Y mutations (501Y.V2 variant) induces conformational change greater than N501Y mutant alone, potentially resulting in an escape mutant. https://doi.org/10.1101/2021.01.13.426558. %J bioRxiv

324. Wang P, Wang M, Yu J, Cerutti G, Nair MS, Huang Y et al (2021) Increased Resistance of SARS-CoV-2 variant P.1 to antibody neutralization. https://doi.org/10.1101/2021.03.01.433466. %J bioRxiv

325. Dejnirattisai W, Zhou D, Supasa P, Liu C, Mentzer AJ, Ginn H et al (2021) Antibody evasion by the Brazilian P.1 strain of SARS-CoV-2. https://doi.org/10.1101/2021.03.12.435194. %J bioRxiv

326. Faria NR, Mellan TA, Whittaker C, Claro IM, da Silva Candido D, Mishra S et al (2021b) Genomics and epidemiology of a novel SARS-CoV-2 lineage in Manaus, Brazil. https://doi.org/10.1101/2021.02.26.21252554. %J medRxiv

327. Freitas ARR, Lemos DRQ, Beckedorff OA, Cavalcante LPG, Siqueira AM, Mello RCS et al (2021) The increase in the risk of severity and fatality rate of covid-19 in southern Brazil after the emergence of the Variant of Concern (VOC) SARS-CoV-2 P.1 was greater among young adults without pre-existing risk conditions. https://doi.org/10.1101/2021.04.13.21255281. %J medRxiv

328. Vogel M, Chang X, Sousa Augusto G, Mohsen MO, Speiser DE, Bachmann MF (2021) SARS-CoV-2 variant with higher affinity to ACE2 shows reduced sera neutralization susceptibility. https://doi.org/10.1101/2021.03.04.433887. %J bioRxiv

329. Montagutelli X, Prot M, Levillayer L, Baquero Salazar E, Jouvion G, Conquet L et al (2021) The B1.351 and P.1 variants extend SARS-CoV-2 host range to mice. https://doi.org/10.1101/2021.03.18.436013. %J bioRxiv

330. Mendes-Correa MC, Vilas-Boas LSS, Bierrenbach AL, Vincente de Paula A, Tozetto-Mendoza TR, Leal FE et al (2021a) Individuals who were mildly symptomatic following infection with SARS-CoV-2 B.1.1.28 have neutralizing antibodies to the P.1 variant. https://doi.org/10.1101/2021.05.11.21256908. %J medRxiv

331. Gidari A, Sabbatini S, Bastianelli S, Pierucci S, Busti C, Monari C et al (2021) Cross-neutralization of SARS-CoV-2 B.1.1.7 and P.1 variants in vaccinated, convalescent and P.1 infected. J Infect. https://doi.org/10.1016/j.jinf.2021.07.019

332. Moreira FRR, D'arc M, Mariani D, Herlinger AL, Schiffler FB, Rossi AD et al (2021) Epidemiological dynamics of SARS-CoV-2 VOC Gamma in Rio de Janeiro, Brazil. https://doi.org/10.1101/2021.07.01.21259404. %J medRxiv

333. Oliveira MdM, Schemberger MO, Suzukawa AA, Riediger IN, Debur MdC, Becker G et al (2021b) Genomic surveillance of SARS-CoV-2 in the state of Paraná, Southern Brazil, reveals the cocirculation of the VOC P.1, P.1-like-II lineage and a P.1 cluster harboring the S:E661D mutation. https://doi.org/10.1101/2021.07.14.21260508. %J medRxiv

334. Franceschi VB, Caldana GD, Perin C, Horn A, Peter C, Cybis GB et al (2021) Predominance of the SARS-CoV-2 lineage P.1 and its sublineage P.1.2 in patients from the metropolitan

region of Porto Alegre, Southern Brazil in March 2021: a phylogenomic analysis. https://doi.org/10.1101/2021.05.18.21257420. %J medRxiv

335. Souza UJB, Santos RN, Belmok A, Melo FL, Galvão JD, Damasceno SB et al (2021) Detection of potential new SARS-CoV-2 Gamma-related lineage in Tocantins shows the spread and ongoing evolution of P.1 in Brazil. https://doi.org/10.1101/2021.06.30.450617. %J bioRxiv

336. Naveca F, Nascimento V, Souza V, Corado A, Nascimento F, Silva G et al (2021b) Emergence and spread of SARS-CoV-2 P.1 (gamma) lineage variants carrying Spike mutations. https://virological.org/t/emergence-and-spread-of-sars-cov-2-p-1-gamma-lineage-variants-carrying-spike-mutations-141-144-n679k-or-p681h-during-persistent-viral-circulation-in-amazonas-brazil/722

337. Resende P, Bezerra J, de Vasconcelos R, Arantes I, Appolinario L, Mendonça A et al (2020)Spike E484K mutation in the first SARS-CoV-2 reinfection case confirmed in Brazil, 2020. https://virological.org/t/spike-e484k-mutation-in-the-first-sars-cov-2-reinfection-case-confirmed-in-brazil-2020/584. Accessed 27 Jan 2021

338. Nonaka C, Miranda Franco M, Gräf TAM, AV, Santana de Aguiar R, Giovanetti M, Solano de Freitas Souza B (2021) Genomic evidence of a Sars-Cov-2 reinfection case with E484K Spike mutation in Brazil. Emerging Infect Dis 27(5). https://doi.org/10.3201/eid2705.210191

339. Yadav P, Mohandas S, Sarkale P, Nyayanit D, Shete A, Sahay R et al (2021a) Isolation of SARS-CoV-2 B.1.1.28.2 P2 variant and pathogenicity comparison with D614G variant in hamster model. https://doi.org/10.1101/2021.05.24.445424. %J bioRxiv

340. Sant'Anna FH, Muterle Varela AP, Prichula J, Comerlato J, Comerlato CB, Roglio VS et al (2021) Emergence of the novel SARS-CoV-2 lineage P.4.1 and massive spread of P.2 in South Brazil. https://doi.org/10.1101/2021.04.14.21255429. %J medRxiv

341. Romero Rebello Moreira F, Menezes Bonfim D, Viana Geddes V, Alves Gomes Zauli D, do Prado Silva J, Brito de Lima A et al (2021) Increasing frequency of SARS-CoV-2 lineages B.1.1.7, P.1 and P.2 and identification of a novel lineage harboring E484Q and N501T spike mutations in Minas Gerais, Southeast Brazil. https://virological.org/t/increasing-frequency-of-sars-cov-2-lineages-b-1-1-7-p-1-and-p-2-and-identification-of-a-novel-lineage-harboring-e484q-and-n501t-spike-mutations-in-minas-gerais-southeast-brazil/676

342. da Silva M, Schons Gularte J, Demoliner M, Witt Hansen A, Heldt F, Filippi M et al (2021) Emergence of a putative novel SARS-CoV-2 P.X lineage harboring N234P and E471Q spike protein mutations in Southern Brazil. Res Square https://doi.org/10.21203/rs.3.rs-689012/v1.

343. Rego N, Salazar C, Paz M, Costábile A, Fajardo A, Ferrés I et al (2021a) Emergence and spread of a B.1.1.28-derived lineage with Q675H and Q677H Spike mutations in Uruguay SARS-CoV-2 coronavirus. https://virological.org/t/emergence-and-spread-of-a-b-1-1-28-derived-lineage-with-q675h-and-q677h-spike-mutations-in-uruguay/730

344. Rego N, Fernández-Calero T, Arantes I, Noya V, Mir D, Brandes M et al (2021b) Spatiotemporal dissemination pattern of SARS-CoV-2 B1.1.28-derived lineages introduced into Uruguay across its southeastern border with Brazil. https://doi.org/10.1101/2021.07.05.21259760. %J medRxiv

345. Ferrareze PAG, Franceschi VB, Mayer AdM, Caldana GD, Zimerman RA, Thompson CE (2021) E484K as an innovative phylogenetic event for viral evolution: Genomic analysis of the E484K spike mutation in SARS-CoV-2 lineages from Brazil. https://doi.org/10.1101/2021.01.27.426895. %J bioRxiv

346. Resende PC, Graf T, Paixao ACD, Appolinario L, Lopes RS, Mendonca ACF et al (2021b) A potential SARS-CoV-2 variant of interest (VOI) harboring mutation E484K in the Spike protein was identified within lineage B.1.1.33 circulating in Brazil. https://doi.org/10.1101/2021.03.12.434969. %J bioRxiv

347. Resende P, Gräf T, Gonçalves Lima Neto L, Vieira da Silva F, Dias Paixão A, Appolinario L et al (2021c) Identification of a new B.1.1.33 SARS-CoV-2 Variant of Interest (VOI) circulating in Brazil with mutation E484K and multiple deletions in the amino (N)-terminal domain of the Spike protein. https://virological.org/t/identification-of-a-new-b-1-1-33-sars-cov-2-variant-of-interest-voi-circulating-in-brazil-with-mutation-e484k-and-multiple-deletions-in-the-amino-n-terminal-domain-of-the-spike-protein/675

348. Duerr R, Dimartino D, Marier C, Zappile P, Wang G, Lighter J et al (2021) Dominance of Alpha and Iota variants in SARS-CoV-2 vaccine breakthrough infections in New York City. https://doi.org/10.1101/2021.07.05.21259547. %J medRxiv

349. West AP, Barnes CO, Yang Z, Bjorkman PJ (2021) SARS-CoV-2 lineage B.1.526 emerging in the New York region detected by software utility created to query the spike mutational landscape. https://doi.org/10.1101/2021.02.14.431043. %J bioRxiv

350. Annavajhala MK, Mohri H, Zucker JE, Sheng Z, Wang P, Gomez-Simmonds A et al (2021) A Novel SARS-CoV-2 variant of concern, B.1.526, identified in New York. https://doi.org/10.1101/2021.02.23.21252259. %J medRxiv

351. Yadav PD, Mohandas S, Shete AM, Nyayanit DA, Gupta N, Patil DY et al (202b) SARS CoV-2 variant B.1.617.1 is highly pathogenic in hamsters than B.1 variant. https://doi.org/10.1101/2021.05.05.442760. %J bioRxiv

352. Baral P, Bhattarai N, Hossen ML, Stebliankin V, Gerstman B, Narasimhan G et al (2021) Mutation-induced changes in the receptor-binding interface of the SARS-CoV-2 delta variant B.1.617.2 and implications for immune evasion. https://doi.org/10.1101/2021.07.17.452576. %J bioRxiv

353. Chaudhari A, Kumar D, Joshi DM, Patel A, Joshi PC (2021) E156/G and Arg158, Phe-157/del mutation in NTD of spike protein in B.1.167.2 lineage of SARS-CoV-2 leads to immune evasion through antibody escape. https://doi.org/10.1101/2021.06.07.447321. %J bioRxiv

354. Arora P, Krueger N, Kempf A, Nehlmeier I, Sidarovich A, Graichen L et al (2021) Increased lung cell entry of B.1.617.2 and evasion of antibodies induced by infection and BNT162b2 vaccination. https://doi.org/10.1101/2021.06.23.449568. %J bioRxiv

355. Hoffmann M, Hofmann-Winkler H, Krueger N, Kempf A, Nehlmeier I, Graichen L et al (2021c) SARS-CoV-2 variant B.1.617 is resistant to Bamlanivimab and evades antibodies induced by infection and vaccination. https://doi.org/10.1101/2021.05.04.442663. %J bioRxiv

356. Ferreira I, Datir R, Papa G, Kemp S, Meng B, Rakshit P et al (2021a) SARS-CoV-2 B.1.617 emergence and sensitivity to vaccine-elicited antibodies. https://doi.org/10.1101/2021.05.08.443253. %J bioRxiv

357. Mohandas S, Yadav PD, Shete AA, Nyayanit D, Sapkal GN, Lole K et al (2021b) SARS-CoV-2 Delta variant pathogenesis and host response in Syrian hamsters. https://doi.org/10.1101/2021.07.24.453631. %J bioRxiv

358. Farinholt T, Qin X, Meng Q, Menon V, Doddapaneni H, Chao H et al (2021) Transmission event of SARS-CoV-2 Delta variant reveals multiple vaccine breakthrough infections. https://doi.org/10.1101/2021.06.28.21258780. %J medRxiv

359. Mishra A, Kumar N, Bhatia S, Aasdev A, Kanniappan S, Thayasekhar A et al (2021) Natural infection of SARS-CoV-2 delta variant in Asiatic lions (Panthera leo persica) in India. https://doi.org/10.1101/2021.07.02.450663. %J bioRxiv

360. Li B, Deng A, Li K, Hu Y, Li Z, Xiong Q et al (2021) Viral infection and Transmission in a large well-traced outbreak caused by the Delta SARS-CoV-2 variant. https://doi.org/10.1101/2021.07.07.21260122. %J medRxiv

361. von Wintersdorff C, Dingemans J, van Alphen L, Wolffs P, van der Veer B, Hoebe C et al (2021) Common laboratory mice are susceptible to infection with SARS-CoV2 beta variant. Res Square. https://doi.org/10.21203/rs.3.rs-777577/v1

362. Tada T, Zhou H, Samanovic MI, Dcosta BM, Cornelius A, Mulligan MJ et al (2021c) Comparison of neutralizing antibody titers elicited by mrna and adenoviral vector vaccine against SARS-CoV-2 variants. https://doi.org/10.1101/2021.07.19.452771. %J bioRxiv

363. Gupta N, Kaur H, Yadav P, Mukhopadhyay L, Sahay RR, Kumar A et al (2021b) Clinical characterization and genomic analysis of COVID-19 breakthrough infections during second wave in different states of India. https://doi.org/10.1101/2021.07.13.21260273. %J medRxiv

364. Kumar V, Singh J, Hasnain SE, Sundar D (2021) Possible link between higher transmissibility of B.1.617 and B.1.1.7 variants of SARS-CoV-2 and increased structural stability of its spike protein and hACE2 affinity. https://doi.org/10.1101/2021.04.29.441933. %J bioRxiv

365. Dhar MS, Marwal R, VS R, Ponnusamy K, Jolly B, Bhoyar RC et al (2021) Genomic characterization and Epidemiology of an emerging SARS-CoV-2 variant in Delhi, India. https://doi.org/10.1101/2021.06.02.21258076. %J medRxiv

366. Kumari D, Kumari N, Kumar S, Sinha PK, Shahi SK, Biswas NR et al (2021) Identification and Characterization of novel mutants of Nsp13 Protein among Indian SARS-CoV-2 isolates. https://doi.org/10.1101/2021.07.30.454406. %J bioRxiv

367. Kimura I, Kosugi Y, Wu J, Yamasoba D, Butlertanaka EP, Tanaka YL et al (2021) SARS-CoV-2 Lambda variant exhibits higher infectivity and immune resistance. https://doi.org/10.1101/2021.07.28.454085. %J bioRxiv

368. Tuccori M, Convertino I, Ferraro S, Valdiserra G, Cappello E, Fini E et al (2021) An overview of the preclinical discovery and development of bamlanivimab for the treatment of novel coronavirus infection (COVID-19): reasons for limited clinical use and lessons for the future. Expert Opin Drug Discov. https://doi.org/10.1080/17460441.2021.1960819

369. Romero P, Dávila-Barclay A, Gonzáles L, Salvatierra G, Cuicapuza D, Solis et al (2021a) Novel sublineage within B.1.1.1 currently expanding in Peru and Chile, with a convergent deletion in the ORF1a gene (Δ3675–3677) and a novel deletion in the Spike gene (Δ246–252, G75V, T76I, L452Q, F490S, T859N). https://virological.org/t/novel-sublineage-within-b-1-1-1-currently-expanding-in-peru-and-chile-with-a-convergent-deletion-in-the-orf1a-gene-3675-3677-and-a-novel-deletion-in-the-spike-gene-246-252-g75v-t76i-l452q-f490s-t859n/685. Accessed 23 Apr 2021

370. Romero PE, Davila-Barclay A, Salvatierra G, Gonzalez L, Cuicapuza D, Solis L et al (2021b) The Emergence of SARS-CoV-2 variant lambda (C.37) in South America. https://doi.org/10.1101/2021.06.26.21259487. %J medRxiv

371. Rose R, Nolan DJ, LaFleur TM, Lamers SL (2021a) Outbreak of P.3 (Theta) SARS-CoV-2 emerging variant of concern among food service workers in Louisiana. https://doi.org/10.1101/2021.07.28.21259040. %J medRxiv

372. Tablizo FA, Kim KM, Lapid CM, Castro MJR, Yangzon MSL, Maralit BA et al (2021) Genome sequencing and analysis of an emergent SARS-CoV-2 variant characterized by multiple spike protein mutations detected from the Central Visayas Region of the Philippines. https://doi.org/10.1101/2021.03.03.21252812. %J medRxiv

373. Bascos NAD, Mirano-Bascos D, Saloma CP (2021) Structural Analysis of Spike Protein Mutations in the SARS-CoV-2 P.3 variant. https://doi.org/10.1101/2021.03.06.434059. %J bioRxiv

374. Hsu S, Parker M, Lindsey B, State A, Zhang P, Foulkes B et al (2021) Detection of Spike mutations; D80G, T95I, G142D, Δ144, N439K, E484K, P681H, I1130V, and D1139H, in B.1.1.482 lineage (AV.1) samples from South Yorkshire, UK

375. Sarkar R, Saha R, Mallick P, Sharma R, Kaur A, Dutta S et al (2021) Emergence of a new SARS-CoV-2 variant from GR clade with a novel S glycoprotein mutation V1230L in West Bengal, India. https://doi.org/10.1101/2021.05.24.21257705. %J medRxiv

376. de Oliveira T, Lutucuta S, Nkengasong J, Morais J, Paula Paixao J, Neto Z et al (2021c) A novel variant of interest of SARS-CoV-2 with multiple spike mutations is identified from travel surveillance in Africa. https://doi.org/10.1101/2021.03.30.21254323. %J medRxiv

377. Choi A, Koch M, Wu K, Dixon G, Oestreicher J, Legault H et al (2021) Serum Neutralizing activity of mRNA-1273 against SARS-CoV-2 variants. https://doi.org/10.1101/2021.06.28.449914. %J bioRxiv

378. Mallm J-P, Bundschuh C, Kim H, Weidner N, Steiger S, Lander I et al (2021) Local emergence and decline of a SARS-CoV-2 variant with mutations L452R and N501Y in the spike protein. https://doi.org/10.1101/2021.04.27.21254849. %J medRxiv

379. Thornlow B, Hinrichs AS, Jain M, Dhillon N, La S, Kapp JD et al (2021) A new SARS-CoV-2 lineage that shares mutations with known Variants of Concern is rejected by automated sequence repository quality control. https://doi.org/10.1101/2021.04.05.438352. %J bioRxiv

380. Ruiz-Rodriguez P, Frances-Gomez C, Chiner-Oms Á, amp, oacutepez MG, Jim et al (2021) Evolutionary and phenotypic characterization of spike mutations in a new SARS-CoV-2 Lineage reveals two Variants of Interest. https://doi.org/10.1101/2021.03.08.21253075. %J medRxiv

381. Mor O, Mandelboim M, Fleishon S, Bucris E, Bar-Ilan D, Linial M et al (2021) The rise and fall of an emerging SARS-CoV-2 variant with the spike protein mutation L452R. https://doi.org/10.1101/2021.07.03.21259957. %J medRxiv

382. Tegally H, Ramuth M, Amoaka D, Scheepers C, Wilkinson E, Giovanetti M et al (2021) A Novel and Expanding SARS-CoV-2 Variant, B.1.1.318, dominates infections in Mauritius. https://doi.org/10.1101/2021.06.16.21259017. %J medRxiv

383. Rodriguez-Maldonado AP, Vazquez-Perez JA, Cedro-Tanda A, Taboada B, Boukadida C, Wong-Arambula C et al (2021) Emergence and spread of the potential variant of interest (VOI) B.1.1.519 predominantly present in Mexico. https://doi.org/10.1101/2021.05.18.212 55620. %J medRxiv

384. Fillatre P, Dufour MJ, Behillil S, Vatan R, Reusse P, Gabellec A et al (2021) A new SARS-CoV-2 variant poorly detected by RT-PCR on nasopharyngeal samples, with high lethality. https://doi.org/10.1101/2021.05.05.21256690. %J medRxiv

385. Tada T, Zhou H, Dcosta BM, Samanovic MI, Mulligan MJ, Landau NR (2021) The Spike proteins of SARS-CoV-2 B.1.617 and B.1.618 variants identified in india provide partial resistance to vaccine-elicited and therapeutic monoclonal antibodies. https://doi.org/10.1101/2021.05.14.444076. %J bioRxiv

386. Dudas G, Hong SL, Potter BI, Calvignac-Spencer S, Niatou-Singa FS, Tombolomako TB et al (2021) Travel-driven emergence and spread of SARS-CoV-2 lineage B.1.620 with multiple VOC-like mutations and deletions in Europe. https://doi.org/10.1101/2021.05.04.21256637. %J medRxiv

387. Laiton-Donato K, Franco-Munoz C, Alvarez-Diaz DA, Ruiz-Moreno H, Usme-Ciro J, Prada D et al (2021b) Characterization of the emerging B.1.621 variant of interest of SARS-CoV-2. https://doi.org/10.1101/2021.05.08.21256619. %J medRxiv

388. Castelli M, Baj A, Criscuolo E, Ferrarese R, Diotti R, Sampaolo M et al (2021) Character-ization of a lineage C.36 SARS-CoV-2 isolate with reduced susceptibility to neutralization circulating in Lombardy, Italy. Viruses

389. Nagano K, Tani-Sassa C, Iwasaki Y, Takatsuki Y, Yuasa S, Takahashi Y et al (2021) SARS-CoV-2 R.1 lineage variants prevailed in Tokyo in March 2021. https://doi.org/10.1101/2021.05.11.21257004. %J medRxiv

390. Prévost J, Richard J, Gasser R, Ding S, Fage C, Anand SP et al (2021) Impact of temperature on the affinity of SARS-CoV-2 Spike for ACE2. https://doi.org/10.1101/2021.07.09.451812. %J bioRxiv

391. Lu Y, Han K, Xue G, Zheng N, Jin G (2021b) Estimated Spike evolution and impact of emerging SARS-CoV-2 variants. https://doi.org/10.1101/2021.05.06.21256705. %J medRxiv

392. Fass E, Zizelski Valenci G, Rubinstein M, Freidlin PJ, Rosencwaig S, Kutikov I et al (2021) HiSpike: a high-throughput cost effective sequencing method for the SARS-CoV-2 spike gene. https://doi.org/10.1101/2021.03.02.21252290. %J medRxiv

393. Vogels C, Alpert T, Breban M, Fauver J, Grubaugh N, Health YSoP: Multiplexed RT-qPCR to screen for SARS-COV-2 B.1.1.7 variants: Preliminary results. https://virological.org/t/mul tiplexed-rt-qpcr-to-screen-for-sars-cov-2-b-1-1-7-variants-preliminary-results/588

394. Korukluoglu G, Kolukirik M, Bayrakdar F, Ozgumus GG, Altas AB, Cosgun Y et al (2021) 40 minutes RT-qPCR assay for screening Spike N501Y and HV69–70del mutations. https://doi.org/10.1101/2021.01.26.428302. %J bioRxiv

395. Wang H, Jean S, Eltringham R, Madison J, Snyder P, Tu H et al (2021e) Mutation-specific SARS-CoV-2 PCR screen: rapid and accurate detection of variants of concern and the iden-tification of a newly emerging variant with Spike L452R mutation. https://doi.org/10.1101/2021.04.22.21255574. %J medRxiv

396. Mattocks CJ, Ward D, Mackay DJ (2021) RT-RC-PCR: a novel and highly scalable next-generation sequencing method for simultaneous detection of SARS-COV-2 and typing variants of concern. https://doi.org/10.1101/2021.03.02.21252704. %J medRxiv

397. Xue T, Wu W, Guo N, Wu C, Huang J, Lai L et al (2020) Single point mutations can potentially enhance infectivity of SARS-CoV-2 revealed by in silico affinity maturation and SPR assay. https://doi.org/10.1101/2020.12.24.424245. %J bioRxiv

398. Chen C, Boorla VS, Banerjee D, Chowdhury R, Cavener VS, Nissly RH et al (2021b) Computational prediction of the effect of amino acid changes on the binding affinity between SARS-CoV-2 spike protein and the human ACE2 receptor. https://doi.org/10.1101/2021.03.24.436885. %J bioRxiv

399. Ortuso F, Mercatelli D, Guzzi PH, Giorgi FM (2020) Structural Genetics of circulating variants affecting the SARS-CoV-2 Spike/human ACE2 complex. https://doi.org/10.1101/2020.09.09.289074. %J bioRxiv

400. Triveri A, Serapian SA, Marchetti F, Doria F, Pavoni S, Cinquini F et al (2021) SARS-CoV-2 Spike protein mutations and escape from antibodies: a computational model of epitope loss in variants of concern. https://doi.org/10.1101/2021.07.12.452002. %J bioRxiv

401. Kim S, Liu Y, Lei Z, Dicker J, Cao Y, Zhang XF et al (2021) Differential interactions between human ACE2 and Spike RBD of SARS-CoV-2 variants of concern. https://doi.org/10.1101/2021.07.23.453598. %J bioRxiv

402. Javanmardi K, Chou C-W, Terrace C, Annapareddy A, Kaoud TS, Guo Q et al (2021) Rapid characterization of spike variants via mammalian cell surface display. https://doi.org/10.1101/2021.03.30.437622. %J bioRxiv

403. Garrett ME, Galloway J, Chu HY, Itell HL, Stoddard CI, Wolf CR et al (2020) High resolution profiling of pathways of escape for SARS-CoV-2 spike-binding antibodies. https://doi.org/10.1101/2020.11.16.385278

404. Shang E, Axelsen P (2020) The potential for SARS-CoV-2 to evade both natural and vaccine-induced immunity. https://doi.org/10.1101/2020.12.13.422567. %J bioRxiv

405. Focosi D, Tuccori M, Baj A, Maggi F (2021) SARS-CoV-2 variants: a synopsis of in vitro efficacy data of convalescent plasma, currently marketed vaccines, and monoclonal antibodies. Viruses 13(7). https://doi.org/10.3390/v13071211

406. Chen J, Gao K, Wang R, Wei G-W (2021c) Revealing the threat of emerging SARS-CoV-2 mutations to antibody therapies. https://doi.org/10.1101/2021.04.12.439473. %J bioRxiv

407. Lusvarghi S, Wang W, Herrup R, Neerukonda SN, Vassell R, Bentley L et al (2021) Key substitutions in the spike protein of SARS-CoV-2 variants can predict resistance to monoclonal antibodies, but other substitutions can modify the effects. https://doi.org/10.1101/2021.07.16.452748. %J bioRxiv

408. Cromer D, Reynaldi A, Steain M, Triccas JA, Davenport MP, Khoury DS (2021) Relating in vitro neutralisation level and protection in the CVnCoV (CUREVAC) trial. https://doi.org/10.1101/2021.06.29.21259504. %J medRxiv

409. Chen X, Chen Z, Azman AS, Sun R, Lu W, Zheng N et al (2021d) Comprehensive mapping of neutralizing antibodies against SARS-CoV-2 variants induced by natural infection or vaccination. https://doi.org/10.1101/2021.05.03.21256506. %J medRxiv

410. Horspool AM, Ye C, Wong TY, Russ BP, Lee KS, Winters MT et al (2021) SARS-CoV-2 B.1.1.7 and B.1.351 variants of concern induce lethal disease in K18-hACE2 transgenic mice despite convalescent plasma therapy. https://doi.org/10.1101/2021.05.05.442784. %J bioRxiv

411. Khoury DS, Cromer D, Reynaldi A, Schlub TE, Wheatley AK, Juno JA et al (2021) Neutralizing antibody levels are highly predictive of immune protection from symptomatic SARS-CoV-2 infection. Nat Med. https://doi.org/10.1038/s41591-021-01377-8

412. Earle KA, Ambrosino DM, Fiore-Gartland A, Goldblatt D, Gilbert PB, Siber GR et al (2021) Evidence for antibody as a protective correlate for COVID-19 vaccines. Vaccine 39(32):4423–4428. https://doi.org/10.1016/j.vaccine.2021.05.063

413. Kustin T, Harel N, Finkel U, Perchik S, Harari S, Tahor M et al (2021) Evidence for increased breakthrough rates of SARS-CoV-2 variants of concern in BNT162b2 mRNA vaccinated individuals. https://doi.org/10.1101/2021.04.06.21254882. %J medRxiv

414. Greaney AJ, Loes AN, Gentles LE, Crawford KH, Starr TN, Malone KD et al (2021c) The SARS-CoV-2 mRNA-1273 vaccine elicits more RBD-focused neutralization, but with broader antibody binding within the RBD. https://doi.org/10.1101/2021.04.14.439844. %J bioRxiv

415. Klingler J, amp, eacuteromine, Lambert GS, Itri V, Liu S et al (2021) SARS-CoV-2 mRNA vaccines induce a greater array of spike-specific antibody isotypes with more potent complement binding capacity than natural infection. https://doi.org/10.1101/2021.05.11.21256972. %J medRxiv

416. Abu Jabal K, Ben Amram H, Beiruti K, Brimat I, Abu Saada A, Bathish Y et alSARS-CoV-2 Immunogenicity in individuals infected before and after COVID-19 vaccination: Israel, January-March 2021: implications for vaccination policy. https://doi.org/10.1101/2021.04.11.21255273. %J medRxiv

417. Ranzani OT, Hitchings M, Dorion Neto M, D'Agostini TL, de Paula RC, de Paula OFP et al (2021) Effectiveness of the CoronaVac vaccine in the elderly population during a P.1 variant-associated epidemic of COVID-19 in Brazil: a test-negative case-control study. https://doi.org/10.1101/2021.05.19.21257472. %J medRxiv

418. Nadesalingam A, Cantoni D, Wells DA, Aguinam ET, Ferrari M, Smith P et al (2021) Breadth of neutralising antibody responses to SARS-CoV-2 variants of concern is augmented by vaccination following prior infection: studies in UK healthcare workers and immunodeficient patients. https://doi.org/10.1101/2021.06.03.21257901. %J medRxiv

419. Shapiro J, Dean NE, Madewell ZJ, Yang Y, Halloran ME, Longini IM (2021) Efficacy estimates for various COVID-19 vaccines: what we know from the literature and reports. https://doi.org/10.1101/2021.05.20.21257461. %J medRxiv

420. Montoya JG, Adams AE, Bonetti V, Deng S, Link NA, Pertsch S et al (2021) Differences in IgG antibody responses following BNT162b2 and mRNA-1273 vaccines. https://doi.org/10.1101/2021.06.18.449086. %J bioRxiv

421. Sokal A, Barba-Spaeth G, Fernandez I, Broketa M, Azzaoui I, de La Selle A et al (2021) Memory B cells control SARS-CoV-2 variants upon mRNA vaccination of naive and COVID-19 recovered individuals. https://doi.org/10.1101/2021.06.17.448459. %J bioRxiv

422. Trinite B, Pradenas E, Marfil S, Rovirosa C, Urrea V, Tarres-Freixas F et al (2021) Previous SARS-CoV-2 infection increases B.1.1.7 cross-neutralization by vaccinated individuals. https://doi.org/10.1101/2021.03.05.433800. %J bioRxiv

423. Leier HC, Bates TA, Lyski ZL, McBride SK, Lee DX, Coulter FJ et al (2021) Previously infected vaccinees broadly neutralize SARS-CoV-2 variants. https://doi.org/10.1101/2021.04.25.21256049. %J medRxiv

424. Lucas C, Vogels CBF, Yildirim I, Rothman J, Lu P, Monteiro V et al (2021) Impact of circulating SARS-CoV-2 variants on mRNA vaccine-induced immunity in uninfected and previously infected individuals. https://doi.org/10.1101/2021.07.14.21260307. %J medRxiv

425. Focosi D, Baj A, Maggi F (2021a) Is a single COVID-19 vaccine dose enough in convalescents? Hum Vaccines Immunother 1–3. https://doi.org/10.1080/21645515.2021.1917238

426. Lustig Y, Nemet I, Kliker L, Zuckerman N, Yishai R, Alroy-Preis S et al (2021) Neutralizing response against variants after SARS-CoV-2 infection and one dose of BNT162b2. 384(25):2453–2454. https://doi.org/10.1056/NEJMc2104036

427. Gallagher KME, Leick MB, Larson RC, Berger TR, Katsis K, Yam JY et al (2021) SARS-CoV-2 T-cell immunity to variants of concern following vaccination. https://doi.org/10.1101/2021.05.03.442455. %J bioRxiv

428. Neidleman J, Luo X, McGregor M, Xie G, Murray V, Greene WC et al (2021) mRNA vaccine-induced SARS-CoV-2-specific T cells recognize B.1.1.7 and B.1.351 variants but differ in longevity and homing properties depending on prior infection status. https://doi.org/10.1101/2021.05.12.443888. %J bioRxiv

429. Luo G, Hu Z, Letterio J (2021) Modeling and predicting antibody durability for mRNA-1273 vaccine for SARS-CoV-2 variants. https://doi.org/10.1101/2021.05.04.21256537. %J medRxiv

430. Pegu A, O'Connell SE, Schmidt SD, O'Dell S, Talana CA, Lai L et al (2021) Durability of mRNA-1273-induced antibodies against SARS-CoV-2 variants. https://doi.org/10.1101/2021.05.13.444010. %J bioRxiv

431. Thomas SJ, Moreira ED, Kitchin N, Absalon J, Gurtman A, Lockhart S et al (2021) Six Month safety and efficacy of the BNT162b2 mRNA COVID-19 vaccine. https://doi.org/10.1101/2021.07.28.21261159. %J medRxiv

432. Bar-On YM, Noor E, Gottlieb N, Sigal A, Milo R (2021) The importance of time post-vaccination in determining the decrease in vaccine efficacy against SARS-CoV-2 variants of concern. https://doi.org/10.1101/2021.06.06.21258429. %J medRxiv

433. Parry HM, Bruton R, Stephens C, Brown K, Amirthalingam G, Hallis B et al (2021) Extended interval BNT162b2 vaccination enhances peak antibody generation in older people. https://doi.org/10.1101/2021.05.15.21257017. %J medRxiv

434. Lin A, Liu J, Ma X, Zhao F, Yu B, He J et al (2021) Heterologous vaccination strategy for containing COVID-19 pandemic. https://doi.org/10.1101/2021.05.17.21257134. %J medRxiv

435. Gram MA, Emborg H-D, Moustsen-Helms IR, Nielsen J, amp, oslashrensen AKB et al (2021) Vaccine effectiveness when combining the ChAdOx1 vaccine as the first dose with an mRNA COVID-19 vaccine as the second dose. https://doi.org/10.1101/2021.07.26.21261130. %J medRxiv

436. Hillus D, Tober-Lau P, Hastor H, Helbig ET, Lippert LJ, Thibeault C et al (2021) Reactogenicity of homologous and heterologous prime-boost immunization with BNT162b2 and ChAdOx1-nCoV19: a prospective cohort study. https://doi.org/10.1101/2021.05.19.212 57334. %J medRxiv

437. Borobia AM, Carcas AJ, Pérez-Olmeda M, Castaño L, Bertran MJ, García-Pérez J et al. Immunogenicity and reactogenicity of BNT162b2 booster in ChAdOx1-S-primed participants (CombiVacS): a multicentre, open-label, randomised, controlled, phase 2 trial. Lancet. https://doi.org/10.1016/S0140-6736(21)01420-3

438. Rose R, Neumann F, Grobe O, Lorentz T, Fickenscher H, Krumbholz A (2021b) Heterologous immunisation with vector vaccine as prime followed by mRNA vaccine as boost leads to humoral immune response against SARS-CoV-2, which is comparable to that according to a homologous mRNA vaccination scheme. https://doi.org/10.1101/2021.07.09.21260251. %J medRxiv

439. Tenbusch M, Schumacher S, Vogel E, Priller A, Held J, Steininger P et al (2021) Heterologous prime-boost vaccination with ChAdOx1 nCoV-19 and BNT162b2 mRNA. https://doi.org/10. 1101/2021.07.03.21258887. %J medRxiv

440. Schmidt T, Klemis V, Schub D, Mihm J, Hielscher F, Marx S et al (2021) Immunogenicity and reactogenicity of a heterologous COVID-19 prime-boost vaccination compared with homologous vaccine regimens. https://doi.org/10.1101/2021.06.13.21258859. %J medRxiv

441. Gross R, Zanoni M, Seidel A, Conzelmann C, Gilg A, Krnavek D et al (2021) Heterologous ChAdOx1 nCoV-19 and BNT162b2 prime-boost vaccination elicits potent neutralizing antibody responses and T cell reactivity. https://doi.org/10.1101/2021.05.30.21257971. %J medRxiv

442. Normark J, Vikström L, Gwon Y-D, Persson I-L, Edin A, Björsell T et al (2021) Heterologous ChAdOx1 nCoV-19 and mRNA-1273 vaccination. https://doi.org/10.1056/NEJMc2110716.

443. Barros-Martins J, Hammerschmidt S, Cossmann A, Odak I, Stankov MV, Morillas Ramos G et al (2021) Humoral and cellular immune response against SARS-CoV-2 variants following heterologous and homologous ChAdOx1 nCoV-19/BNT162b2 vaccination. https://doi.org/10.1101/2021.06.01.21258172. %J medRxiv

444. Roozen GVT, Prins M, van Binnendijk R, den G, Kuiper V, Prins C et al (2021) Tolerability, safety and immunogenicity of intradermal delivery of a fractional dose mRNA-1273 SARS-CoV-2 vaccine in healthy adults as a dose sparing strategy. https://doi.org/10.1101/2021.07.27.21261116. %J medRxiv

445. Mohamed SS, Chao GYC, Isho B, Zuo M, Nahass GR, Salomon-Shulman RE et al (2021) A mucosal antibody response is induced by intra-muscular SARS-CoV-2 mRNA vaccination. https://doi.org/10.1101/2021.08.01.21261297. %J medRxiv

446. Riemersma KK, Grogan BE, Kita-Yarbro A, Jeppson GE, O'Connor DH, Friedrich TC et al (2021) Vaccinated and unvaccinated individuals have similar viral loads in communities with a high prevalence of the SARS-CoV-2 delta variant. https://doi.org/10.1101/2021.07.31.212 61387. %J medRxiv

447. Van Egeren D, Novokhodko A, Stoddard M, Tran U, Zetter B, Rogers M et al (2020) Risk of evolutionary escape from neutralizing antibodies targeting SARS-CoV-2 spike protein. https://doi.org/10.1101/2020.11.17.20233726. %J medRxiv

448. Landis J, Moorad R, Pluta LJ, Caro-Vegas C, McNamara RP, Eason AB et al (2021) Intra-host evolution provides for continuous emergence of SARS-CoV-2 variants. https://doi.org/10.1101/2021.05.08.21256775. %J medRxiv

449. Kabinger F, Stiller C, Schmitzová J, Dienemann C, Hillen HS, Höbartner C et al (2021) Mechanism of molnupiravir-induced SARS-CoV-2 mutagenesis. https://doi.org/10.1101/2021.05.11.443555. %J bioRxiv

450. Andreano E, Piccini G, Licastro D, Casalino L, Johnson NV, Paciello I et al (2020) SARS-CoV-2 escape in vitro from a highly neutralizing COVID-19 convalescent plasma. https://doi.org/10.1101/2020.12.28.424451. %J bioRxiv

451. Valesano AL, Rumfelt KE, Dimcheff DE, Blair CN, Fitzsimmons WJ, Petrie JG et al (2021) Temporal dynamics of SARS-CoV-2 mutation accumulation within and across infected hosts. https://doi.org/10.1101/2021.01.19.427330. %J bioRxiv

452. Choi B, Choudhary MC, Regan J, Sparks JA, Padera RF, Qiu X et al (2020) Persistence and Evolution of SARS-CoV-2 in an Immunocompromised Host. 383(23):2291–2293. https://doi.org/10.1056/NEJMc2031364

453. Dong J, Zost SJ, Greaney AJ, Starr TN, Dingens AS, Chen EC et al (2021) Genetic and structural basis for recognition of SARS-CoV-2 spike protein by a two-antibody cocktail. https://doi.org/10.1101/2021.01.27.428529. %J bioRxiv

454. Lohr B, Niemann D, Verheyen J (2021) Bamlanivimab treatment leads to rapid selection of immune escape variant carrying E484K mutation in a B.1.1.7 infected and immunosuppressed patient. Clin Infect Dis: Off Publ Infect Dis Soc Am. https://doi.org/10.1093/cid/ciab392

455. Focosi D, Novazzi F, Genoni A, Dentali F, Dalla Gasperina D, Baj A et alDaniele Focosi, Federica Novazzi, Angelo Genoni, Francesco Dentali, Daniela Dalla gasperina, Andreina Baj, Fabrizio Maggi. 2021.

456. Aydillo T, Gonzalez-Reiche AS, Aslam S, van de Guchte A, Khan Z, Obla A et al (2020) Shedding of viable SARS-CoV-2 after immunosuppressive therapy for cancer. 383(26):2586–2588. https://doi.org/10.1056/NEJMc2031670

457. Avanzato VA, Matson MJ, Seifert SN, Pryce R, Williamson BN, Anzick SL et al (2020) Case study: prolonged infectious SARS-CoV-2 shedding from an asymptomatic immunocompromised individual with cancer. Cell. https://doi.org/10.1016/j.cell.2020.10.049

458. Hensley MK, Bain WG, Jacobs J, Nambulli S, Parikh U, Cillo A et al (2021) Intractable COVID-19 and prolonged SARS-CoV-2 replication in a CAR-T-cell therapy recipient: a case study. Clin Infect Dis: off Publ Infect Dis Soc Am. https://doi.org/10.1093/cid/ciab072

459. Kemp SA, Collier DA, Datir R, Gayed S, Jahun A, Hosmillo M et al (2020b) Neutralising antibodies in Spike mediated SARS-CoV-2 adaptation. Nature https://doi.org/10.1101/2020.12.05.20241927. %J medRxiv

460. Chen L, Zody MC, Mediavilla JR, Cunningham MH, Composto K, Chow KF et al (2021e) Emergence of multiple SARS-CoV-2 antibody escape variants in an immunocompromised host undergoing convalescent plasma treatment. https://doi.org/10.1101/2021.04.08.21254791. %J medRxiv

461. Monrad I, Sahlertz SR, Nielsen SSF, Pedersen L, Petersen MS, Kobel CM et al (2021) Persistent severe acute respiratory syndrome coronavirus 2 infection in immunocompromised host displaying treatment induced viral evolution. Open Forum Infect Dis 8(7):ofab295. https://doi.org/10.1093/ofid/ofab295

462. Bazykin G, Stanevich O, Danilenko D, Fadeev A, Komissarova K, Ivanova A et al (202)Emergence of Y453F and Δ69–70HV mutations in a lymphoma patient with long-term COVID-19. https://virological.org/t/emergence-of-y453f-and-69-70hv-mutations-in-a-lymphoma-patient-with-long-term-covid-19/580

463. Borges V, Isidro J, Cunha M, Cochicho D, Martins L, Banha L et al (2021b) Long-term evolution of SARS-CoV-2 in an immunocompromised patient with non-Hodgkin lymphoma. mSphere. https://doi.org/10.1128/mSphere.00244-21

464. Karim F, Moosa MY, Gosnell B, Sandile C, Giandhari J, Pillay S et al (2021) Persistent SARS-CoV-2 infection and intra-host evolution in association with advanced HIV infection. https://doi.org/10.1101/2021.06.03.21258228. %J medRxiv

465. Kavanagh Williamson M, Hamilton F, Hutchings S, Pymont HM, Hackett M, Arnold D et al (2021) Chronic SARS-CoV-2 infection and viral evolution in a hypogammaglobulinaemic individual. https://doi.org/10.1101/2021.05.31.21257591. %J medRxiv

466. Mendes-Correa MC, Ghilardi F, Salomao MC, Villas-Boas LS, Vincente de Paula A, Tozetto-Mendoza TR et al (2021b) SARS-CoV-2 shedding, infectivity and evolution in an immunocompromised adult patient. https://doi.org/10.1101/2021.06.11.21257717. %J medRxiv

467. Roquebert B, Trombert-Paolantoni S, Haim-Boukobza S, Lecorche E, Verdurme L, Foulonge V et al (2021) The SARS-CoV-2 B.1.351/V2 variant is outgrowing the B.1.1.7/V1 variant in French regions in April 2021. https://doi.org/10.1101/2021.05.12.21257130. %J medRxiv

468. Yang W, Shaman J (2021) Epidemiological characteristics of three SARS-CoV-2 variants of concern and implications for future COVID-19 pandemic outcomes. https://doi.org/10.1101/2021.05.19.21257476. %J medRxiv

469. Bolze A, Cirulli ET, Luo S, White S, Cassens T, Jacobs S et al (2021) Rapid displacement of SARS-CoV-2 variant B.1.1.7 by B.1.617.2 and P.1 in the United States. https://doi.org/10.1101/2021.06.20.21259195. %J medRxiv

470. WHO (2021) Weekly epidemiological update on COVID-19—27 Apr 2021

471. Pung R, Mak TM, Kucharski AJ, Lee VJ (2021) Serial intervals observed in SARS-CoV-2 B.1.617.2 variant cases. https://doi.org/10.1101/2021.06.04.21258205. %J medRxiv

472. Dagpunar JS (2021) Interim estimates of increased transmissibility, growth rate, and reproduction number of the Covid-19 B.1.617.2 variant of concern in the United Kingdom. https://doi.org/10.1101/2021.06.03.21258293. %J medRxiv

473. Challen R, Dyson L, Overton C, Guzman-Rincon L, Hill E, Stage H et al (2021) Early epidemiological signatures of novel SARS-CoV-2 variants: establishment of B.1.617.2 in England. https://doi.org/10.1101/2021.06.05.21258365. %J medRxiv

474. Ito K, Piantham C, Nishiura H (2021) Predicted domination of variant Delta of SARS-CoV-2 before Tokyo Olympic games, Japan. https://doi.org/10.1101/2021.06.12.21258835. %J medRxiv

475. Duchene S, Featherstone L, Haritopoulou-Sinanidou M, Rambaut A, Lemey P, Baele G (2020) Temporal signal and the phylodynamic threshold of SARS-CoV-2. Virus Evol 6(2). https://doi.org/10.1093/ve/veaa061

476. Redd AD, Nardin A, Kared H, Bloch EM, Pekosz A, Laeyendecker O et al (2021) CD8+ T cell responses in COVID-19 convalescent individuals target conserved epitopes from multiple prominent SARS-CoV-2 circulating variants. https://doi.org/10.1101/2021.02.11.21251585. %J medRxiv

477. Eguia R, Crawford KHD, Stevens-Ayers T, Kelnhofer-Millevolte L, Greninger AL, Englund JA et al (2020) A human coronavirus evolves antigenically to escape antibody immunity. https://doi.org/10.1101/2020.12.17.423313. %J bioRxiv

478. Bewick S (2021) Viral variants and vaccinations: if we can change the COVID-19 vaccine, should we? https://doi.org/10.1101/2021.01.05.21249255. %J medRxiv

479. Baum A, Fulton BO, Wloga E, Copin R, Pascal KE, Russo V et al (2020) Antibody cocktail to SARS-CoV-2 spike protein prevents rapid mutational escape seen with individual antibodies. 369(6506):1014–1018. https://doi.org/10.1126/science.abd0831 %J Science

480. Focosi D, Tuccori M, Antonelli G, Maggi F (2020) What is the optimal usage of Covid-19 convalescent plasma donations? Clin Microb Infect. S1198-743X(20):30589–30599

481. Focosi D, Maggi F (2021) Neutralising antibody escape of SARS-CoV-2 spike protein: risk assessment for antibody-based Covid-19 therapeutics and vaccines. Rev Med Virol. https://doi.org/10.1002/rmv.2231

482. Montefiori DC, Acharya P (2021) SnapShot: SARS-CoV-2 antibodies. Cell Host Microbe 29(7):1162–.e1. https://doi.org/10.1016/j.chom.2021.06.005

483. Finkelstein MT, Mermelstein AG, Parker Miller E, Seth PC, Stancofski ED, Fera D (2021) Structural analysis of neutralizing epitopes of the SARS-CoV-2 Spike to guide therapy and vaccine design strategies. Viruses 13(1). https://doi.org/10.3390/v13010134

484. Rakha A, Rasheed H, Batool Z, Akram J, Adnan A, Du J (2020) COVID-19 variants database: a repository for human SARS-CoV-2 polymorphism data. https://doi.org/10.1101/2020.06.10.145292. %J bioRxiv

485. Xing Y, Li X, Gao X, Dong Q (2020) MicroGMT: a mutation tracker for SARS-CoV-2 and other microbial genome sequences. 11(1502). https://doi.org/10.3389/fmicb.2020.01502

486. Mercatelli D, Triboli L, Fornasari E, Ray F, Giorgi FM. Coronapp: a web application to annotate and monitor SARS-CoV-2 mutations.n/a(n/a). https://doi.org/10.1002/jmv.26678

487. De Silva NH, Bhai J, Chakiachvili M, Contreras-Moreira B, Cummins C, Frankish A et al (2020) The Ensembl COVID-19 resource: ongoing integration of public SARS-CoV-2 data. https://doi.org/10.1101/2020.12.18.422865. %J bioRxiv

488. Desai SS, Rane A, Joshi A, Dutt A (2021) Evolving insights from SARS-CoV-2 genome from 200K COVID-19 patients. https://doi.org/10.1101/2021.01.21.427574. %J bioRxiv

489. Tzou PL, Tao K, Nouhin J, Rhee S-Y, Hu BD, Pai S et al (2020) Coronavirus antiviral research database (CoV-RDB): an online database designed to facilitate comparisons between candidate anti-coronavirus compounds. 12(9):1006

490. Ford CT, Scott R, Jacob Machado D, Janies D. Sequencing data of North American SARS-CoV-2 isolates shows widespread complex variants. https://doi.org/10.1101/2021.01.27.212 50648. %J medRxiv

491. Wittig A, Miranda F, Tang M, Hölzer M, Renard BY, Fuchs S. CovRadar: Continuously tracking and filtering SARS-CoV-2 mutations for molecular surveillance. https://doi.org/10. 1101/2021.02.03.429146. %J bioRxiv

492. Lucaci A, Zehr J, Shank S, Bouvier D, Mei H, Nekrutenko A et al (2021) RASCL: rapid assessment of SARS-COV-2 clades enabled through molecular sequence analysis and its application to B.1.617.1 and B.1.617.2. https://virological.org/t/rascl-rapid-assessment-of-sars-cov-2-clades-enabled-through-molecular-sequence-analysis-and-its-application-to-b-1-617-1-and-b-1-617-2/709

493. Fang S, Li K, Shen J, Liu S, Liu J, Yang L et al (2020) GESS: a database of global evaluation of SARS-CoV-2/hCoV-19 sequences. Nucl Acids Res. https://doi.org/10.1093/nar/gkaa808

494. Ferreira R-C, Wong E, Gugan G, Wade K, Liu M, Munoz Baena L et al (2021b) CoVizu: rapid analysis and visualization of the global diversity of SARS-CoV-2 genomes. https://doi.org/10.1101/2021.07.20.453079. %J bioRxiv

495. Haag E, Latif A, Gangavarapu K, Mullen J, Tsueng G, Matteson N et al (2021) outbreak.info: SARS-CoV-2 mutation situation reports. https://virological.org/t/outbreak-info-sars-cov-2-mutation-situation-reports/629 Accessed 27 Feb 2021

496. Pucci F, Rooman M (2021) Prediction and evolution of the molecular fitness of SARS-CoV-2 variants: introducing SpikePro. 4.11.439322. https://doi.org/10.1101/2021.04.11.439322. %J bioRxiv

497. Wagner C, Hodcroft E, Bell S, Neher R, Bedford T (2021) Resurgence of SARS-CoV-2 19B clade corresponds with possible convergent evolution. https://nextstrain.org/groups/blab/nar ratives/ncov/19B/2020-02-16. Accessed 16 Feb 2021

498. Murall C, Mostefai F, Grenier J, Poujol R, Hussin J, Moreira S et al (2021) Recent evolution and international transmission of SARS-CoV-2 clade 19B (Pango A lineages). https://virological.org/t/recent-evolution-and-international-transmission-of-sars-cov-2-clade-19b-pango-a-lineages/711. Accessed 6 June 2021.

499. Pater AA, Bosmeny MS, Barkau CL, Ovington KN, Chilamkurthy R, Parasrampuria M et al (2021) Emergence and evolution of a prevalent New SARS-CoV-2 variant in the United States. https://doi.org/10.1101/2021.01.11.426287. %J bioRxiv

500. Tomkins-Tinch C, Silbert J, DeRuff K, Siddle K, Gladden-Young A, Bronson AMJ et al (2021) Detection of the recurrent substitution Q677H in the spike protein of SARS-CoV-2 in cases descended from the lineage B.1.429. https://virological.org/t/detection-of-the-recurr ent-substitution-q677h-in-the-spike-protein-of-sars-cov-2-in-cases-descended-from-the-lin eage-b-1-429/660

501. Shen L, Dien Bard J, Triche TJ, Judkins AR, Biegel JA, Gai X (2021) Emerging variants of concern in SARS-CoV-2 membrane protein: a highly conserved target with potential pathological and therapeutic implications. https://doi.org/10.1101/2021.03.11.434758. %J bioRxiv

502. Bates TA, Leier HC, Lyski ZL, McBride SK, Coulter FJ, Weinstein JB et al (2021) Neutralization of SARS-CoV-2 variants by convalescent and vaccinated serum. https://doi.org/10.1101/2021.04.04.21254881. %J medRxiv

503. Fenwick C, Turelli P, Pellaton C, Farina A, Campos J, Raclot C et al (2021) A multiplexed high-throughput neutralization assay reveals a lack of activity against multiple variants after SARS-CoV-2 infection. https://doi.org/10.1101/2021.04.08.21255150. %J medRxiv

504. Caniels TG, Bontjer I, van der Straten K, Poniman M, Burger JA, Appelman B et al (2021) Emerging SARS-CoV-2 variants of concern evade humoral immune responses from infection and vaccination. https://doi.org/10.1101/2021.05.26.21257441. %J medRxiv

505. Dupont L, Snell LB, Graham C, Seow J, Merrick B, Lechmere T et al (2021) Antibody longevity and cross-neutralizing activity following SARS-CoV-2 wave 1 and B.1.1.7 infections. https://doi.org/10.1101/2021.06.07.21258351. %J medRxiv

506. Kaku Y, Kuwata T, Zahid HM, Hashiguchi T, Noda T, Kuramoto N et al (2021) Resistance of SARS-CoV-2 variants to neutralization by antibodies induced in convalescent patients with COVID-19. Cell Rep. https://doi.org/10.1016/j.celrep.2021.109385

507. Vacharathit V, Aiewsakun P, Manopwisedjaroen S, Srisaowakarn C, Laopanupong T, Ludowyke N et al (2021) SARS-CoV-2 variants of concern exhibit reduced sensitivity to live-virus neutralization in sera from CoronaVac vaccinees and naturally infected COVID-19 patients. https://doi.org/10.1101/2021.07.10.21260232. %J medRxiv

508. Müller L, Andree M, Moskorz W, Drexler I, Walotka L, Grothmann R et al (2021b) Age-dependent immune response to the Biontech/Pfizer BNT162b2 COVID-19 vaccination. https://doi.org/10.1101/2021.03.03.21251066. %J medRxiv

509. Yadav PD, Sapkal G, Ella R, Sahay RR, Nyayanit DA, Patil DY et al (2021c) Neutralization against B.1.351 and B.1.617.2 with sera of COVID-19 recovered cases and vaccinees of BBV152. https://doi.org/10.1101/2021.06.05.447177. %J bioRxiv

510. Shen X, Tang H, Pajon R, Smith G, Glenn GM, Shi W et al (2021) Neutralization of SARS-CoV-2 Variants B.1.429 and B.1.351. 384(24):2352–2354. https://doi.org/10.1056/NEJMc2103740

511. Yadav P, Sapkal GN, Abraham P, Ella R, Deshpande G, Patil DY et al (2021d) Neutralization of variant under investigation B.1.617 with sera of BBV152 vaccinees. https://doi.org/10.1101/2021.04.23.441101. %J bioRxiv

512. Planas D, Veyer D, Baidaliuk A, Staropoli I, Guivel-Benhassine F, Rajah M et al (2021b) Reduced sensitivity of SARS-CoV-2 variant Delta to antibody neutralization. Nature. https://doi.org/10.1038/s41586-021-03777-9

513. Sapkal G, Yadav PD, Ella R, Abraham P, Patil DY, Gupta N et al (2021b) Neutralization of B.1.1.28 P2 variant with sera of natural SARS-CoV-2 infection and recipients of BBV152 vaccine. https://doi.org/10.1101/2021.04.30.441559. %J bioRxiv

514. Zhou H, Dcosta BM, Samanovic MI, Mulligan MJ, Landau NR, Tada T (2021) B.1.526 SARS-CoV-2 variants identified in New York City are neutralized by vaccine-elicited and therapeutic monoclonal antibodies. https://doi.org/10.1101/2021.03.24.436620. %J bioRxiv

515. Tang J, Lee Y, Ravichandran S, Grubbs G, Huang C, Stauft C et al (2021) Reduced neutralization of SARS-CoV-2 variants by convalescent plasma and hyperimmune intravenous immunoglobulins for treatment of COVID-19. https://doi.org/10.1101/2021.03.19.436183. %J bioRxiv

516. Zhang L, Huynh T, Luan B (2021e) *In silico* assessment of antibody drug resistance to Bamlanivimab of SARS-CoV-2 Variant B.1.617. https://doi.org/10.1101/2021.05.12.443826. %J bioRxiv

517. Westendorf K, Žentelis S, Foster D, Vaillancourt P, Wiggin M, Lovett E et al (2021) LY-CoV1404 potently neutralizes SARS-CoV-2 variants. https://doi.org/10.1101/2021.04.30.442182. %J bioRxiv

518. Fenwick C, Turelli P, Pellaton C, Farina A, Campos J, Raclot C et al (2021b) A high-throughput cell- and virus-free assay shows reduced neutralization of SARS-CoV-2 variants by COVID-19 convalescent plasma. https://doi.org/10.1126/scitranslmed.abi8452. %J Science Translational Medicine

519. Lee S-Y, Ryu D-K, Choi YK, Moore P, Baalen CAV, Song R et al (2021a) Therapeutic effect of CT-P59 against SARS-CoV-2 South African variant. https://doi.org/10.1101/2021.04.27. 441707. %J bioRxiv

520. Lee S-Y, Ryu D-K, Kang B, Noh H, Kim J, Seo J-M et al (2021b) Therapeutic efficacy of CT-P59 against P.1 variant of SARS-CoV-2. https://doi.org/10.1101/2021.07.08.451696. %J bioRxiv

521. Ryu D-K, Woo H-M, Kang B, Noh H, Kim J-I, Seo J-M et al (2021) The in vitro and in vivo potency of CT-P59 against Delta and its associated variants of SARS-CoV-2. https://doi.org/10.1101/2021.07.23.453472. %J bioRxiv

522. Cathcart AL, Havenar-Daughton C, Lempp FA, Ma D, Schmid M, Agostini ML et al (2021) The dual function monoclonal antibodies VIR-7831 and VIR-7832 demonstrate potent in vitro and in vivo activity against SARS-CoV-2. https://doi.org/10.1101/2021.03.09.434607. %J bioRxiv

523. Lanini S, Milleri S, Andreano E, Nosari S, Paciello I, Piccini G et al (2021) A single intramuscular injection of monoclonal antibody MAD0004J08 induces in healthy adults SARS-CoV-2 neutralising antibody titres exceeding those induced by infection and vaccination. https://doi.org/10.1101/2021.08.03.21261441. %J medRxiv

524. Andreano E, Nicastri E, Paciello I, Pileri P, Manganaro N, Piccini G et al (2021) Extremely potent human monoclonal antibodies from COVID-19 convalescent patients. Cell 184(7):1821–35.e16. https://doi.org/10.1016/j.cell.2021.02.035

525. Becker M, Dulovic A, Junker D, Ruetalo N, Kaiser P, Pinilla Y et al (2021) Immune response to SARS-CoV-2 variants of concern in vaccinated individuals. https://doi.org/10.1101/2021.03.08.21252958. %J medRxiv

526. Gonzalez C, Saade C, Bal A, Valette M, Saker K, Lina B et al (2021) Live virus neutralization testing in convalescent and vaccinated subjects against 19A, 20B, 20I/501Y.V1 and 20H/501.V2 isolates of SARS-CoV-2. https://doi.org/10.1101/2021.05.11.21256578. %J medRxiv

527. Liu J, Bodnar BH, Wang X, Wang P, Meng F, Khan AI et al (2021c) Correlation of vaccine-elicited antibody levels and neutralizing activities against SARS-CoV-2 and its variants. https://doi.org/10.1101/2021.05.31.445871. %J bioRxiv

528. Garcia-Beltran WF, Lam EC, Denis KS, Nitido AD, Garcia ZH, Hauser BM et al (2021) Multiple SARS-CoV-2 variants escape neutralization by vaccine-induced humoral immunity. https://doi.org/10.1101/2021.02.14.21251704. %J medRxiv

529. Wall EC, Wu M, Harvey R, Kelly G, Warchal S, Sawyer C et al (2021) Neutralising antibody activity against SARS-CoV-2 VOCs B.1.617.2 and B.1.351 by BNT162b2 vaccination. Lancet. 397(10292):2331–2333. https://doi.org/10.1016/S0140-6736(21)01290-3

530. Stankov M, Cossmann A, Bonifacius A, Jablonka A, Morillas Ramos G, Goedecke N et al (2021) Humoral and cellular immune responses against SARS-CoV-2 variants and human coronaviruses after single BNT162b2 vaccination. https://doi.org/10.1101/2021.04.16.21255412. %J medRxiv

531. Strengert M, Becker M, Ramos GM, Dulovic A, Gruber J, Juengling J et al (2021) Cellular and humoral immunogenicity of a SARS-CoV-2 mRNA vaccine in patients on hemodialysis. https://doi.org/10.1101/2021.05.26.21257860. %J medRxiv

532. Abu-Raddad LJ, Chemaitelly H, Butt AA (2021b) Effectiveness of the BNT162b2 Covid-19 Vaccine against the B.1.1.7 and B.1.351 Variants. https://doi.org/10.1056/NEJMc2104974

533. Rovida F, Cassaniti I, Paolucci S, Percivalle E, Sarasini A, Piralla A et al (2021) SARS-CoV-2 vaccine breakthrough infections are asymptomatic or mildly symptomatic and are infrequently transmitted. https://doi.org/10.1101/2021.06.29.21259500. %J medRxiv

534. Lopez Bernal J, Andrews N, Gower C, Gallagher E, Simmons R, Thelwall S et al (2021) Effectiveness of COVID-19 vaccines against the B.1.617.2 variant. New England J Med. https://doi.org/10.1056/NEJMoa2108891

535. Edara V-V, Lai L, Sahoo M, Floyd K, Sibai M, Solis D et al (2021c) Infection and vaccine-induced neutralizing antibody responses to the SARS-CoV-2 B.1.617.1 variant. https://doi.org/10.1101/2021.05.09.443299. %J bioRxiv

536. Liu J, Liu Y, Xia H, Zou J, Weaver S, Swanson KA et al (2021d) BNT162b2-elicited neutral-
     ization of B.1.617 and other SARS-CoV-2 variants. Nature. https://doi.org/10.1038/s41586-
     021-03693-y
537. Edara V-V, Pinsky BA, Suthar MS, Lai L, Davis-Gardner ME, Floyd K et alI (2021) nfection
     and vaccine-induced neutralizing-antibody responses to the SARS-CoV-2 B.1.617 Variants.
     New England J Med. https://doi.org/10.1056/NEJMc2107799
538. Sheikh A, McMenamin J, Taylor B, Robertson C. SARS-CoV-2 Delta VOC in Scotland:
     demographics, risk of hospital admission, and vaccine effectiveness. Lancet. https://doi.org/
     10.1016/S0140-6736(21)01358-1
539. Shen X, Tang H, McDana C, Wagh K, Fischer W, Theiler J et al (2021) SARS-CoV-2 variant
     B.1.1.7 is susceptible to neutralizing antibodies elicited by ancestral Spike vaccines. Cell Host
     Microbe 29(4):529–539. https://doi.org/10.1101/2021.01.27.428516. %J bioRxiv
540. Corbett KS, Werner A, O'Connell S, Gagne M, Lai L, Moliva J et al (2021) Evaluation of
     mRNA-1273 against SARS-CoV-2 B.1.351 Infection in Nonhuman Primates. https://doi.org/
     10.1101/2021.05.21.445189. %J bioRxiv
541. Edara VV, Hudson WH, Xie X, Ahmed R, Suthar MS (2021e) Neutralizing antibodies against
     SARS-CoV-2 variants after infection and vaccination. JAMA. https://doi.org/10.1001/jama.
     2021.4388%JJAMA
542. Fischer R, van Doremalen N, Adney D, Yinda C, Port J, Holbrook M et al (2021) ChAdOx1
     nCoV-19 (AZD1222) protects against SARS-CoV-2 B.1.351 and B.1.1.7. https://doi.org/10.
     1101/2021.03.11.435000. %J bioRxiv
543. Yadav P, Sapkal GN, Abraham P, Deshpande G, Nyayanit D, Patil DY et al (2021e) Neutral-
     ization potential of Covishield vaccinated individuals against B.1.617.1. https://doi.org/10.
     1101/2021.05.12.443645. %J bioRxiv
544. Sapkal GN, Yadav PD, Sahay RR, Deshpande G, Gupta N, Nyayanit DA et al (2021c) Neutral-
     ization of Delta variant with sera of Covishield vaccinees and COVID-19 recovered vaccinated
     individuals. https://doi.org/10.1101/2021.07.01.450676. %J bioRxiv
545. Kale P, Gupta E, Bihari C, Patel N, Rooge S, Pandey A et al (2021) Clinicogenomic analysis
     of breakthrough infections by SARS CoV2 variants after ChAdOx1 nCoV-19 vaccination in
     healthcare workers. https://doi.org/10.1101/2021.06.28.21259546. %J medRxiv
546. Spencer AJ, Morris S, Ulaszewska M, Powers C, Kaliath R, Bissett CD et al (2021) The
     ChAdOx1 vectored vaccine, AZD2816, induces strong immunogenicity against SARS-CoV-
     2 B.1.351 and other variants of concern in preclinical studies. https://doi.org/10.1101/2021.
     06.08.447308. %J bioRxiv
547. Moore P, Moyo T, Hermanus T, Kgagudi P, Ayres F, Makhado Z et al (2021) Neutralizing
     antibodies elicited by the Ad26.COV2.S COVID-19 vaccine show reduced activity against
     501Y.V2 (B.1.351), despite protection against severe disease by this variant. https://doi.org/
     10.1101/2021.06.09.447722. %J bioRxiv
548. Jongeneelen M, Kaszas K, Veldman D, Huizingh J, van der Vlugt R, Schouten T et al (2021)
     Ad26.COV2.S elicited neutralizing activity against Delta and other SARS-CoV-2 variants of
     concern. https://doi.org/10.1101/2021.07.01.450707. %J bioRxiv
549. Ikegame S, Siddiquey MNA, Hung C-T, Haas G, Brambilla L, Oguntuyo KY et al (2021)
     Qualitatively distinct modes of Sputnik V vaccine-neutralization escape by SARS-CoV-2
     Spike variants. https://doi.org/10.1101/2021.03.31.21254660. %J medRxiv
550. Shinde V, Bhikha S, Hossain Z, Archary M, Bhorat Q, Fairlie L et al (2021) Efficacy of NVX-
     CoV2373 Covid-19 Vaccine against the B.1.351 Variant. New England J Med 384(20):1899–
     1909. https://doi.org/10.1101/2021.02.25.21252477. %J medRxiv
551. Yadav PD, Sahay RR, Sapkal G, Nyayanit D, Shete A, Deshpande G et al (2021) Comparable
     neutralization of SARS-CoV-2 Delta AY.1 and Delta in individuals sera vaccinated with
     BBV152. https://doi.org/10.1101/2021.07.30.454511. %J bioRxiv
552. Ella R, Vadrevu KM, Jogdand H, Prasad S, Reddy S, Sarangi V et al (2021) Safety and immuno-
     genicity of an inactivated SARS-CoV-2 vaccine, BBV152: a double-blind, randomised, phase
     1 trial. Lancet Infect Dis 21(5):637–646. https://doi.org/10.1016/S1473-3099(20)30942-7

553. Chen Y, Shen H, Huang R, Tong X, Wu C. Serum neutralising activity against SARS-CoV-2 variants elicited by CoronaVac. Lancet Infect Dis. https://doi.org/10.1016/S1473-3099(21)00287-5

554. Acevedo ML, Alonso-Palomares L, Bustamante A, Gaggero A, Paredes F, Cortés CP et al (2021) Infectivity and immune escape of the new SARS-CoV-2 variant of interest Lambda. https://doi.org/10.1101/2021.06.28.21259673. %J medRxiv

555. Hu J, Wei X-y, Xiang J, Peng P, Xu F-l, Wu K et al (2021) Reduced neutralization of SARS-CoV-2 B.1.617 variant by inactivated and RBD-subunit vaccine. https://doi.org/10.1101/2021.07.09.451732. %J bioRxiv

Printed in the United States
by Baker & Taylor Publisher Services